Springer
Tokyo
Berlin
Heidelberg
New York
Barcelona
Hong Kong
London
Milan
Paris
Singapore

Jinkan Sai, Joe Ariyama

MRCP
Early Diagnosis of
Pancreatobiliary Diseases

 Springer

Jinkan Sai, M.D.
Department of Gastroenterology
Juntendo University
2-1-1 Hongo, Bunkyo-ku, Tokyo 113-8421, Japan

Joe Ariyama, M.D.
Professor, Department of Gastroenterology
Juntendo University
2-1-1 Hongo, Bunkyo-ku, Tokyo 113-8421, Japan

ISBN 978-4-431-70273-3 Springer-Verlag Tokyo Berlin Heidelberg New York

Printed on acid-free paper

© Springer-Verlag Tokyo 2000
This English translation is based on the Japanese original, J. Sai; Magnetic Resonance
Cholangiopancreatography
Published by Chugai Igaku
© 1998 Chugai Igaku

SPIN: 10743846

Foreword

This book serves as a comprehensive and up-to-date guide to magnetic resonance cholangiopancreatography (MRCP). MRCP is a newly developed technique which is noninvasive and does not require administration of contrast materials. MRCP is an application of MR imaging that provides both high-quality cross-sectional images of ductal structures and projectional images of the pancreatobiliary ducts. The technique is based on a heavily T2-weighted sequence, as a result of which stationary fluid including bile and pancreatic juice has a high signal intensity, while solid organs have a low signal intensity. Due to rapid evolution of the technology, high- quality 2-D or 3-D MRCP images are now available.

Since the introduction of MRCP, there has been a significant decrease in the use of diagnostic endoscopic retrograde cholangiopancreatography (ERCP) at our institution. MRCP is now indicated following ultrasound and computed tomography in patients with suspected pancreatobiliary diseases. MRCP has routinely replaced diagnostic ERCP because of high sensitivity and specificity of MRCP in the diagnosis of pancreatobiliary diseases.

I would like to express my considerable gratitude to Professor Hitoshi Katayama, Dr. Masahiro Irimoto, and Chief Radiographer Eiju Itsumi and other radiographers in the Department of Radiology; to Mr. Junichi Makita of the MR Engineering Department of Toshiba Corporation; to Assistant Professor Masafumi Suyama and staff members of the Department of Gastroenterology; to Professor Shunji Futakawa and Associate Professor Tomoe Beppu of the 2nd Department of Surgery; and to Professor Koichi Suda of the Department of Pathology.

We hope that this book may serve many readers and we look forward to receiving constructive criticism.

October 1999
Joe Ariyama, M.D.
Professor of Gastroenterology
Juntendo University School of Medicine
Tokyo

Preface

Magnetic resonance cholangiopancreatography (MRCP) is a truly innovative noninvasive diagnostic technique which permits the visualization of the anatomy and pathology of the pancreatobiliary tree. The major benefit for patients is that MRCP enables the early diagnosis of pancreatobiliary diseases in a noninvasive manner, and the results obtained are directly applicable to a treatment plan.

Based on our clinical experience of 3000 MRCP studies, we are convinced of the clinical usefulness of MRCP for the diagnosis of a variety of pancreatobiliary diseases. This book presents many clinical cases to illustrate the effectiveness of this new diagnostic modality. It is our hope that this book will help to promote the acceptance of MRCP in routine clinical practice.

I dedicate this book to my parents, Sang-U Sai and Yang-Ja Son; to my wife, Hae-Ran Kim; and to my children, Sung-Kyu and Myeung-Hwa, for their support, understanding, and encouragement.

October 1999
Jinkan Sai

Contents

Chapter 1 Concept of Magnetic Resonance Cholangiopancreatography

1.1 Introduction to Magnetic Resonance Cholangiopancreatography

Magnetic resonance cholangiopancreatography (MRCP) is a new imaging technique for visualizing the biliary tract and pancreatic duct using the principle of magnetic resonance imaging. Using heavily T2-weighted images to visualize only stationary fluid such as bile and pancreatic juice, this technique enables the visualization of the morphology of the pancreaticobiliary tree [1, 2].

This innovative technique is noninvasive, requires no injection of contrast medium, and offers the unique ability to acquire both projectional and cross-sectional images of the ductal system [3, 4]. Since MRCP has high sensitivity to water, it provides more information than the cholangiopancreatographic images obtained by endoscopic retrograde cholangiopancreatography (ERCP) (Table 1-1). As a result, a better understanding of a pancreatobiliary disease and its detailed pathology can be obtained, which enables accurate diagnosis and formulation of an appropriate treatment plan. Therefore, MRCP is expected to replace diagnostic ERCP for the screening of pancreatobiliary diseases [5, 6].

MRCP has gained popularity rapidly due to its usefulness in routine clinical practice, and is becoming an indispensable diagnostic modality for pancreatobiliary diseases.

Table 1-1 Information acquired from MRCP, which is superior to ERCP

The biliary tract and pancreatic duct both proximal and distal to a site of obstruction	Fig. 3-12, 13, 14, 29, 30, 31, Fig. 4-13, 25
The gallbladder in cases of cystic duct obstruction	Fig. 3-10, 16, 34, 44
Inflammatory effusion in cases of cholangitis or pancreatitis	Fig. 3-16, 17; Fig. 4-4
Visualizing cysts with no communication with the biliary tract and pancreatic duct	Fig. 4-6, 7, 13, 14, 22, 24, Fig. 5-3, 4
Internal structure of cystic lesions from review of source images	Fig. 4-16, 19, 22, Fig. 5-4
Visualizing the biliary tract and pancreatic duct in cases of mucin-producing tumor, which is not satisfactorily imaged by ERCP due to the presence of mucin	Fig. 4-16, 17, 19
Cholangiopancreatography after reconstruction of the upper gastrointestinal tract	Fig. 3-39, 40, 41
Rokitansky-Aschoff sinus (RAS) of the gallbladder	Fig. 3-20, 21, 22, 23
Patency of the stent	Fig. 3-45
Functional image of the pancreas	Fig. 6-1, 2
Cystic lesions in the liver and kidney	Fig. 5-1, 2
Liver abscess with no communication with the biliary duct	Fig. 5-8

1.2 Comparison of MRCP with Conventional Cholangiopancreatic Imaging Techniques

MRCP offers various advantages over conventional cholangiopancreatic imaging techniques such as intravenous cholangiography, percutaneous transhepatic cholangiography, computed tomographic (CT) cholangiography, and endoscopic retrograde cholangiopancreatography (ERCP) [6-8].

In terms of safety and comfort to patients, MRCP requires no injection of contrast medium or ionizing radiation and involves no pain or complication [9,10]. It can be performed even during the acute phase of pancreatitis and cholangitis, and the result is not affected by serum bilirubin level as is intravenous cholangiography or CT cholangiography [11]. It provides more information about the pancreatobiliary pathology than ERCP (Table 1-1) and can be performed in patients who present difficulties with ERCP examination, such as those who have undergone biliary enteric anastomosis or those with stenosis of the upper gastrointestinal tract preventing passage of the endoscope. It can also be performed in patients with poor clinical performance status. A higher success rate is therefore achieved. Technically, MRCP does not depend on the operator's skill. There is also no need for premedication before examination [5].

On the other hand, MRCP has some disadvantages [5]. For example, spatial resolution is lower than ERCP. And while biopsy, cytology, or therapeutic intervention can be performed with ERCP, they are not possible with MRCP. Overlapping of fluid collection or cystic structure may disturb the visualization of the pancreatobiliary tree. Direct visualization of the papilla of Vater is impossible. It is also difficult to confirm the patency of the cystic duct or the communication between cysts and the ductal system. In addition, the image quality is deteriorated by artifacts derived from some types of surgical clips, coils, or metallic stents in the abdomen. MRCP is contraindicated in patients having certain types of implants such as pacemakers and some types of intracranial clips and heart valves. And MRCP cannot be used to study patients with claustrophobia.

1.3 Role and Indications of MRCP

MRCP has been found to have wide clinical indications due to its ability to depict the physiologic and pathological state of the pancreatic duct and biliary tract. An important role of MRCP is in replacing diagnostic ERCP as a noninvasive modality [8, 12]. Unsuccessful or incomplete ERCP cases are satisfactorily examined by MRCP [13]. Furthermore, MRCP provides detailed information on the pancreaticobiliary tree not obtainable by tomographic imaging techniques such as ultrasonography (US) and CT. Because MRCP can be conducted conveniently and safely in the outpatient clinic, it is extremely useful as a screening (Table 1-2) and follow-up method for pancreatobiliary diseases [6], although its cost effectiveness and accessibility remain as problems.

Table 1-2 Screening for pancreatobiliary diseases using MRCP

Anatomic variants

      e.g., pancreaticobiliary maljunction (Figs. 3-6, 7), pancreatic divisum (Fig. 4-3), low cystic duct insertion (Fig. 3-3), aberrant right hepatic duct (Fig. 3-2)

Stones

      e.g., gallstones (Fig. 3-9–16), pancreatolithiasis (Figs. 4-5, 8, 9)

Cystic disease

      e.g., choledochal cyst (Fig. 3-7), pancreatic cyst (Figs. 4-13, 14, 22–24), intraductal papillary mucinous tumor of the pancreas (Fig. 4-16–21), Von Meyenburg complex (Fig. 5-3)

Ductal stenosis

      e.g., carcinoma of the biliary tract (Figs. 3-25, 26, 27, 29), Mirizzi's syndrome (Fig. 3-34), primary sclerosing cholangitis (Fig. 3-35), postoperative bile duct stenosis (Fig. 3-38), carcinoma of the papilla of Vater (Fig. 3-31, 33), benign stenosis of the papilla of Vater (Fig. 3-37), pancreatic carcinoma (Fig. 4-25, 29, 30), chronic pancreatitis (Fig. 4-7, 10, 11, 13)

Elevated lesions

      e.g., carcinoma of the biliary tract (Fig. 3-28), carcinoma of the papilla of Vater (Fig. 3-30, 32), intraductal papillary tumor of the pancreas (Fig. 4-16, 19)

Acute inflammation

      e.g., acute cholecystitis (Fig. 3-16, 17), acute pancreatitis (Fig. 4-4)

Others

      e.g., adenomyomatosis of the gallbladder (Fig. 3-20–23), biliary sludge (Fig. 3-43, 44)

Clinical Indications for MRCP

1. Early diagnosis of pancreatobiliary malignancy
2. Evaluation of pancreatobiliary emergency, such as obstructive jaundice, acute pancreatitis, and acute cholecystitis
3. Examination of the biliary tract before cholecystectomy to detect choledocholithiasis or anatomic variants of the biliary tree
4. Survey after biliary enteric anastomosis
5. Follow-up study of pancreatobiliary diseases including intraductal papillary mucinous tumor of the pancreas, pancreatitis, pancreatic cyst, benign stricture of the biliary tract, and patency of biliary stent
6. Cases in which ERCP is contraindicated

References

1. Wallner B, Schumacher K, Friedrich J (1991) Dilated biliary tract: Evaluation with MR cholangiography with a T2-weighted contrast-enhanced fast sequence. Radiology 181:805-808

2. Takehara Y, Ichijo K, Tooyama N, et al (1994) Breath-hold MR cholangiography with a long-echo-train fast spin-echo sequence and a surface coil in chronic pancreatitis. Radiology 192:73-78

3. Morimoto K, Shimoi M, Shirakawa T, et al (1992) Biliary obstruction: evaluation with three-dimensional MR cholangiography. Radiology 183:578-580

4. Soto JA, Barish MA, Yucel EK, et al (1995) Pancreatic duct: MR cholangio-pancreatography with a three-dimensional fast spin-echo technique. Radiology 196:459-464

5. Reinhold C, Bret PM (1996) Current status of MR cholangiopancreatography. AJR Am J Roentgenol 166:1285-1295

6. Sai J, Ariyama J (1999) MRCP in the diagnosis of pancreatobiliary diseases: its progression and limitation (in Japanese). Jpn J Gastroenterol 96:259-265

7. Reuther G, Kiefer B, Tuchmann A (1996) Cholangiography before biliary surgery: single shot MR cholangiopancreatography versus intravenous cholangiography. Radiology 198:561-566

8. Takehara Y (1998) Can MRCP replace ERCP? J Magn Reson Imaging 8:517-534

9. Bilbao MK, Dotter CT, Lee TG, et al (1976) Complications of endoscopic retrograde cholangiopancreatography (ERCP). A study of 10,000 cases. Gastroenterology 70:314-320

10. Ariyama J (1988) Percutaneous transhepatic cholangiography. In: Margulis AR, Burenne HJ (eds) Alimentary tract radiology. Mosby, St. Louis, pp 2229-2241

11. Stockberger SM, Wass JL, Sherman S, et al (1994) Intravenous cholangiography with helical CT: comparison with endoscopic retrograde cholangiography. Radiology 192: 675-680

12. Fulcher AS, Turner MA, Capps GW, et al (1998) Half-Fourier RARE MR cholangiopancreatography: experience in 300 subjects. Radiology 207:21-32

13. Soto JA, Yucel EK, Barish MA, et al (1996) MR cholangiopancreatography after unsuccessful or incomplete ERCP. Radiology 199:91-98

Chapter 2 Imaging Techniques of MRCP

2.1 Technical Development

2.1.1   Gradient Echo Sequence

In 1991, Wallner et al. [1] and Morimoto et al. [2] first reported MR cholangiography using a steady-state free procession (SSFP) gradient echo sequence. However, the sequence required long period of breath holding, and the spatial resolution was inadequate to allow satisfactory imaging of the pancreatic duct. The reason was that the gradient echo sequence was susceptible to magnetic field inhomogeneity and motions such as gastrointestinal peristalsis, pulsation of the aorta, and abdominal wall movement.

2.1.2   Spin Echo Sequence

2.1.2.1   Fast Spin Echo Technique

In 1994, Takehara et al.[3] proposed the fast spin echo (FSE) technique with a long echo train that yielded heavily T2-weighted MRCP images with higher spatial resolution than the SSFP sequence. The FSE technique had a higher signal-to-noise and contrast-to-noise ratio, and was less susceptible to motion and magnetic field inhomogeneity. Furthermore, advances in the MRI system, such as the introduction of phased-array multicoils, improved the spatial resolution. Application of a fat saturation technique diminished the background signal due to intraabdominal fat, contributing to better quality of the MRCP images. Consequently, satisfactory images of the pancreatobiliary tree were acquired with the FSE technique [4], although routine visualization of the nondilated pancreatic duct was still difficult. The major disadvantage of the FSE technique was that it required a long acquisition time and a breath-holding scan was difficult to acquire in most patients. As a result, image quality was deteriorated by respiratory motion artifacts [5]. Another drawback of FSE technique was that the sequence required maximum intensity projection (MIP) for postprocessing, which caused a high false-positive rate of ductal stenosis due to misregistration artifacts [6].

2.1.2.2   Single-Shot Rapid Acquisition by Relaxation Enhancement (RARE) Technique

In 1995, Laubenberger et al. [7] proposed the single-shot rapid acquisition by relaxation enhancement (RARE) technique, which made it possible to acquire all the echo signals (240 echoes) from a single 90°

excitation pulse with an effective echo time (TE) of 1100 ms. RARE had a considerably high signal-to-noise ratio for a long T2 component of water-containing structure, while signal sources other than fluids were strongly suppressed.

In this technique, MRCP was acquired during a single breath-hold of 4 s, acquisition of the entire field of view was possible using a single thick slice, and a projectional image was acquired without postprocessing by MIP. As a result, the spatial resolution was improved and continuous images of the nondilated pancreatobiliary tract was acquired routinely without artifacts arising from respiration or MIP reconstruction. The single-slice technique allowed imaging in any arbitrary plane, although the MIP technique had the advantage of allowing visualization at any angle of choice.

2.1.2.3   Half-Fourier RARE Technique

In 1995, Hirohashi et al. [8] described MRCP using a half-Fourier acquisition single-shot turbo spin-echo (HASTE) technique. This was a single-shot FSE technique employing improved Fourier transformation of RARE. This technique acquired only half the lines in k space, permitting reduction of the scanning time to 2 s. Subsequently, other modifications such as fast asymmetric spin echo (FASE) [9] and single-shot fast spin echo (SSFSE) [10] were introduced.

We introduced the FASE technique in 1995, and all images in this book were obtained using this technique.

2.2 Imaging Procedures

2.2.1 Patient Preparation

In principle, patients undergoing MRCP do not require any special preparation. However, to ensure the best possible results, patients should be fasted for 4 to 6 h before the examination. Administration of an antiperistaltic drug such as butyl-scopolamine is useful to minimize artifacts caused by fast bowel movement, although this is not always required in the single-shot FSE techniques such as RARE, HASTE, SSFSE, and FASE, because the scanning time is only 2 to 4 s [11, 12].

2.2.2 Patient Position

Patients are usually examined in the supine position. When a spine coil is used, the prone position generally provides better images [13].

2.2.3 Breathing Methods

Respiratory motion artifacts cause significant deterioration of image quality of MRCP. The single-shot FSE techniques (RARE, HASTE, SSFSE, and FASE) permit scanning during a single breath-hold of only 2 to 4 s, and images obtained by these techniques are therefore free from respiratory motion artifacts [14, 15]. On the other hand, the conventional FSE technique requires a long scanning time, and respiratory triggering, quit breathing, intermittent breath-holding, or a long period of breath-holding with oxygen inhalation has to be used [3,5,16].

2.2.4 Receiver Coils

Surface coils increase the signal-to-noise ratio and improve the spatial resolution of MRCP. Phased array coils [4] have become the standard in recent years.

2.2.5   Scanning Parameters

The optimal scanning parameters differ depending on the imaging system. The scanning conditions for FASE sequence used at our institution are shown in Table 2-1.

2.2.6   Scanning Planes

In the single-shot FSE techniques (RARE, HASTE, SSFSE, and FASE), a single thick slice is scanned in the coronal plane, left anterior oblique (LAO) plane, right anterior oblique (RAO) plane, and axial plane (Fig. 2-1). Acquisition with any arbitrary plane is feasible, and the optimal plane to visualize the pancreatobiliary pathology can be acquired. Furthermore multi-sections in the coronal or axial plane are acquired followed by postprocessing with maximum intensity projection (MIP), which is capable of reconstructing three-dimensional (3-D) images viewed from any desired angle.

2.2.7   Maximum Intensity Projection

Maximum intensity projection (MIP) is a postprocessing technique to reconstruct a 3-D projection image from a series of cross-sectional images. This method displays the maximum intensity voxel encountered in the projected ray, and projection images from various viewing angles can be reconstructed [17].

2.2.8   Fat Saturation

Fat tissue in the abdomen increases the background signal of MRCP images and deteriorates the image quality, unless the effective TE is longer than 1000 ms. Therefore, fat saturation is required to acquire a clear MRCP image [12]. In our institution the frequency-selective fat saturation method is used.

2.2.9   Administration of Ferric Ammonium Citrate

In spin echo sequences, the gastrointestinal fluid produces high signal intensity and may obscure the MRCP images. Oral administration of a concentrated solution of ferric ammonium citrate serves as a negative

Table 2-1 FASE sequence

Effective echo time	250 ms
Echo train length	212
Band width	167 kHz
Echo spacing	12.5 msec
Slice thickness	2D: 40 mm
	3D: 1.5 mm $\times$ 35 to 45
Matrix	384 $\times$ 384
FOV	35 $\times$ 35 cm

Fig.2-1 Scanning planes for a single thick slice of two-dimensional MRCP using FASE technique

a. Coronal plane: to visualize the pancreatic duct in the body of the pancreas and common bile duct.
b. Right anterior oblique plane: to visualize the pancreatic duct in the tail of the pancreas and the gallbladder.
c. Left anterior oblique plane: to visualize the pancreatic duct in the head of the pancreas, the lower bile duct, and the pancreaticobiliary junction.

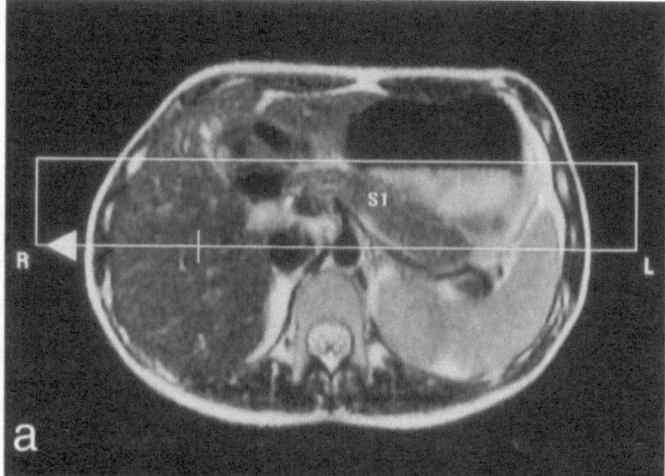

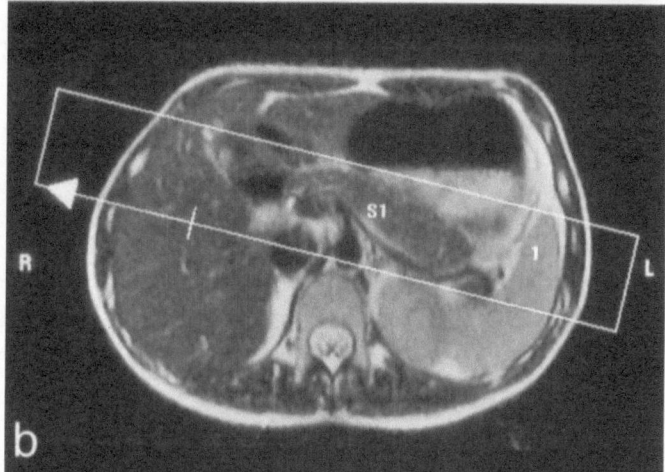

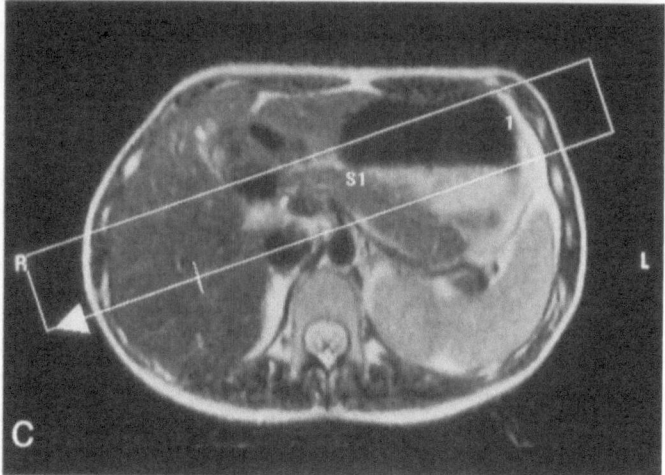

Fig. 2-2 Two-dimensional MRCP before and after administration of ferric ammonium citrate

a. Before administration

High signal intensity of the gastric and duodenal fluid obscure the pancreaticobiliary tree.

b. After administration

Gastric and duodenal fluid have resulted in low signal intensity, and the pancreaticobiliary tree is clearly visible.

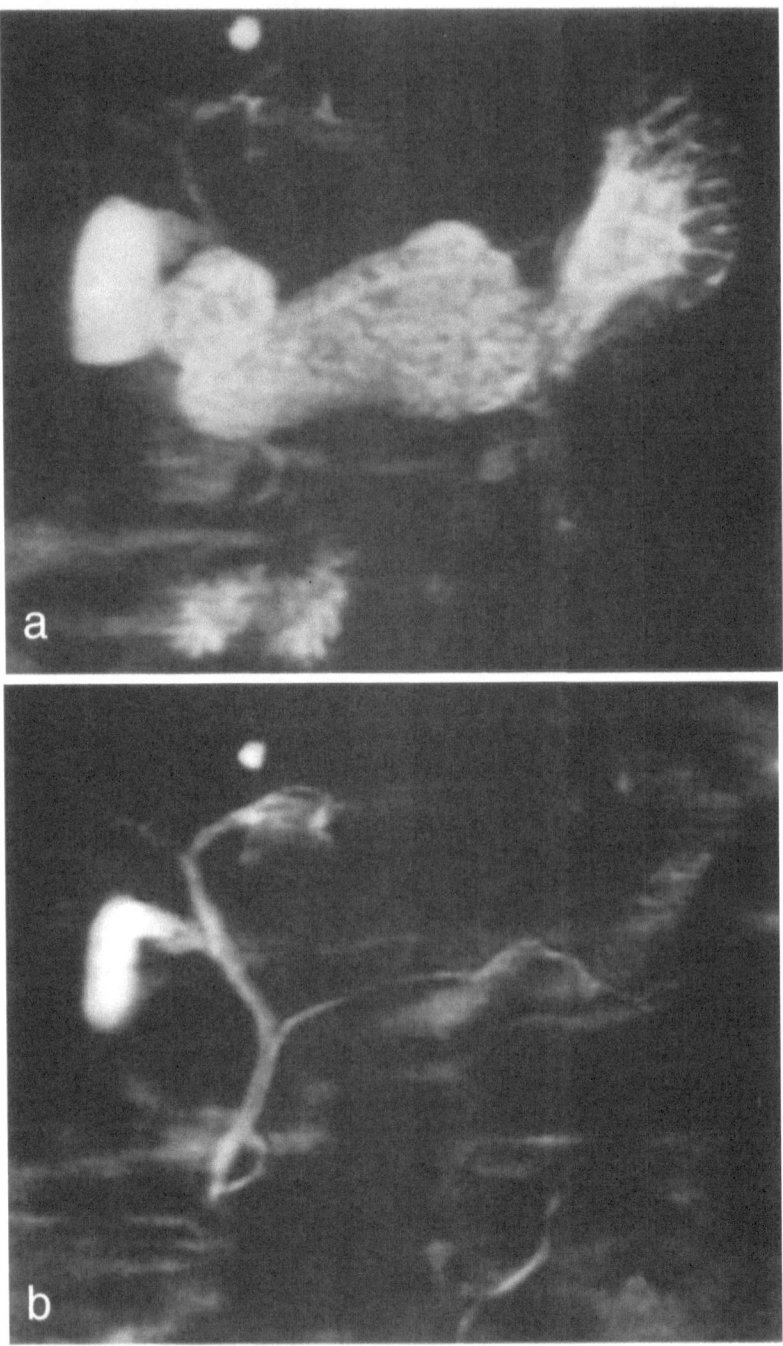

contrast agent for T2-weighted images and improves the image quality by eliminating the signals from gastrointestinal fluid that may overlap with the MRCP images [18, 19] (Fig. 2-2). Usually, 1200 mg of ferric ammonium citrate is dissolved in 70 ml of water and given orally just before the examination.

2.2.10   Intravenous Injection of Secretin

Intravenous injection of secretin is useful to delineate the pancreatic duct clearly [20,21], because the pancreatic duct is not always filled with pancreatic juice. Secretin increases the absolute volume of intraductal free water by stimulating the secretion of pancreatic juice. Usually, 50–100 U of secretin is injected intravenously.

2.3 Two-dimensional versus three-dimensional images

Two-dimensional (2-D) imaging using single-shot FSE techniques (RARE, HASTE, SSFSE, and FASE) acquires a single thick slice and requires only a brief breath-hold of 2 to 4 s. This short acquisition time minimizes artifacts caused by respiratory motions and gastrointestinal peristalsis, and images with high spatial resolution can be obtained. Furthermore, the scanning plane can be selected freely. Image reconstruction does not involve any special procedure such as MIP [7, 8].

On the other hand, 3-D acquisition with MIP achieves a higher signal-to-noise ratio (SNR) and contrast-to-noise ratio (CNR) than 2-D acquisition (Fig. 3-8) and yields satisfactory MRCP images even in lower magnetic field. The image can be rotated at any angle (Fig. 3-31, 4-16, 4-22). By reviewing the source images, resolution of the intraductal pathology, such as stones or polypoid lesions, is superior to 2-D acquisition, because 2-D acquisition with a single thick slice may suffer from partial volume effects (Fig. 3-8) [12, 22].

Although a 2-D image is inferior to a 3-D image in terms of CNR, SNR, and resolution of the intraductal pathology, it is simple and quick, less susceptible to artifacts, and is therefore more suitable for routine clinical practice including the examination of elderly patients or those in poor physical condition [13].

References

1. Wallner B, Schumacher K, Friedrich J (1991) Dilated biliary tract: Evaluation with MR cholangiography with a T2-weighted contrast-enhanced fast sequence. Radiology 181:805-808

2. Morimoto K, Shimoi M, Shirakawa T, et al (1992) Biliary obstruction: evaluation with three-dimensional MR cholangiography. Radiology 183:578-580

3. Takehara Y, Ichijo K, Tooyama N, et al (1994) Breath-hold MR cholangiography with a long-echo-train fast spin-echo sequence and a surface coil in chronic pancreatitis. Radiology 192:73-78

4. Bret PM, Reinhold C, Taourel P, et al (1996) Pancreas divisum: evaluation with MR cholangiopancreatography. Radiology 199:99-103

5. Guibaud L, Bret PM, Reinhold C, et al (1995) Bile duct obstruction and choledocholithiasis: diagnosis with MR cholangiopancreatography. Radiology 197:109-115

6. Anderson CM, Saloner D, Tsuruda JS, Shapeero LG, Lee RE (1990) Artifacts in maximum-intensity-projection display of MR angiography. AJR Am J Roentgenol 154:623-629

7. Laubenberger J., Buchert M, Shneider B, et al (1995) Breath-hold projection magnetic resonance cholangiopancreatography (MRCP): a new method for the examination of the bile duct and pancreatic ducts. Magn Reson Med 33:18-23

8. Hirohashi S, Kitano S, Hirohashi R, et al (1995) Evaluation of pancreaticobiliary system using MR cholangiopancreatography (abstr) (in Japanese). Nippon Acta Radiol 55:S74

9. Sai J, Ariyama J, Suyama M, et al (1996) Breath-hold MR cholangiopancreatography with fast advanced spin echo technique in the diagnosis of malignant obstruction of the lower biliary tract. HPB Surg 9:433

10. Takehara Y (1999) Fast MR imaging for evaluating the pancreaticobiliary system. Eur J Radiol 29:211-232

11. Sai J, Ariyama J (1999) MRCP in the diagnosis of pancreatobiliary diseases: its progression and limitation (in Japanese). Jpn J Gastroenterol 96:259-265

12. Takehara Y (1998) Can MRCP replace ERCP? JMRI 8:517-534

13. Sai J, Ariyama J, Suyama M, et al (1999) MR cholangiopancreatography of the normal pancreatic duct system. Med Rev 69:27-31

14. Reuther G, Kiefer B, Tuchmann A (1996) Cholangiography before biliary surgery: single shot MR cholangiopancreatography versus intravenous cholangiography. Radiology 198:561-566

15. Miyazaki K, Yamashita Y, Tsuchigame T, et al (1996) MR cholangiopancreatography using HASTE (half-Fourier acquisition single-shot turbo spin-echo) sequences. AJR Am J Roentgenol 166:1297-1303

16. Soto JA, Barish MA, Yucel EK, et al (1995) Pancreatic duct: MR cholangio-pancreatography with a three-dimensional fast spin-echo technique. Radiology 196:459-464

17. Rossnick S, Laub G, Braeckle R, et al (1986) Three-dimensional display of blood vessels in MRI. In: Proceedings of the IEEE Computers in Cardiology Conference. Institute of Electrical and Electric Engineers, New York, pp 193-196

18. Takahara T, Saeki M, Nosaka S, et al (1995) The use of high concentration ferric ammonium citrate (FAC) solution as a negative bowel contrast agent: application in MR cholangiopancreatography (in Japanese). Nippon Acta Radiol 55:697-699

19. Hirohashi S, Hirohashi R, Uchida H, et al (1997) MR cholangiopancreatography and MR urography : improved enhancement with a negative oral contrast agent. Radiology 203:281-285

20. Takehara Y, Ichijo K, Tooyama N, et al (1995) Enhanced delineation of the pancreatic duct in MR cholangiopancreatography (MRCP) with a combined use of secretin (in Japanese). Nippon Acta Radiol 55:255-256

21. Matos C, Metens T, Deviere J, et al (1997) Pancreatic duct: morphologic and functional evaluation with dynamic MR pancreatography after secretin stimulation. Radiology 203:435-441

22. Reinhold C, Bret PM (1996) Current status of MR cholangiopancreatography. AJR Am J Roentgenol 166:1285-1295

Chapter 3 Biliary Tract

3.1 Normal Biliary Tract

MRCP using the single-shot fast spin echo technique (RARE, HASTE, FASE, SSFSE) allows routine visualization of the normal biliary tract, including the first to third level branches of the intrahepatic bile duct, left and right hepatic ducts, and the common hepatic and common bile duct [1, 2].

In our study of 50 normal subjects, MRCP depicted the normal biliary tract at a rate of 98% for the right hepatic duct, 100% for the left hepatic duct, 100% for the common bile duct, 92% for the cystic duct, 100% for the gallbladder, and 96% for the pancreatobiliary junction [3]. (Fig. 3-1).

MRCP depicts the physiological state of the biliary tract. The duct caliber on ERCP may be larger than that seen on MRCP, due to the pressure of the injected contrast material in ERCP.

Fig. 3-1   Normal biliary tract

Single-slice MRCP depicts the normal biliary tract, including the first to third level branches of the intrahepatic bile duct, left and right hepatic duct, and the common hepatic and common bile duct. The cystic duct is also clearly demonstrated.

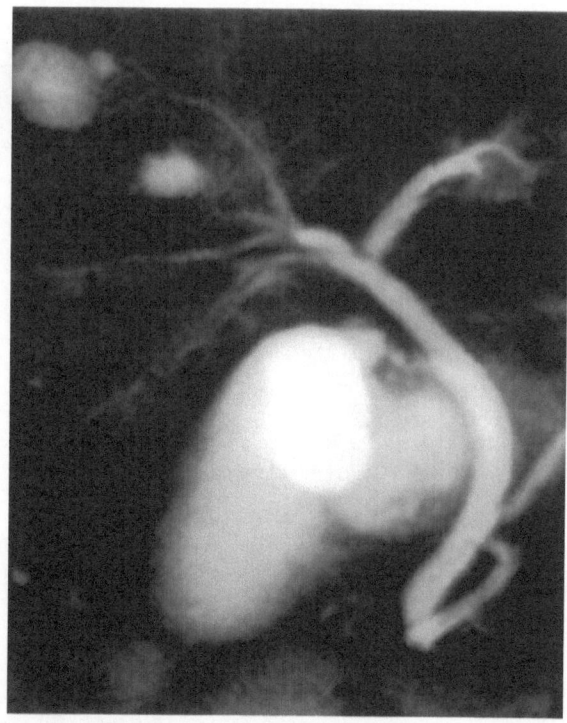

3.2 Anatomic Variants of the Biliary Tree

MRCP is useful for evaluating anatomic variants of the biliary tree that may increase the risk of bile duct injury during laparoscopic cholecystectomy [4,5]. An aberrant right hepatic duct, a low and medial cystic duct insertion, and a parallel course of the cystic and hepatic duct have been reported at a prevalence of 6.5%, 10%, 18%, and 11%, respectively [6,7]. The rate of biliary tract injury associated with laparoscopic cholecystectomy is reported as 0.2–0.3% [8].

In our study of 122 patients before cholecystectomy, the accuracy of MRCP in the diagnosis of an aberrant right hepatic duct, a low or medial cystic duct insertion, and a parallel course of the cystic duct and hepatic duct were 100%, 93%, 96%, and 88% respectively [9].

Fig. 3-2   Aberrant right hepatic duct (*arrow*)

Single-slice MRCP shows joining of the right anterior hepatic duct and the left hepatic duct at the level of the bile duct bifurcation, and the right posterior hepatic duct (*arrow*) directly drains into the common hepatic duct (*arrowhead*).

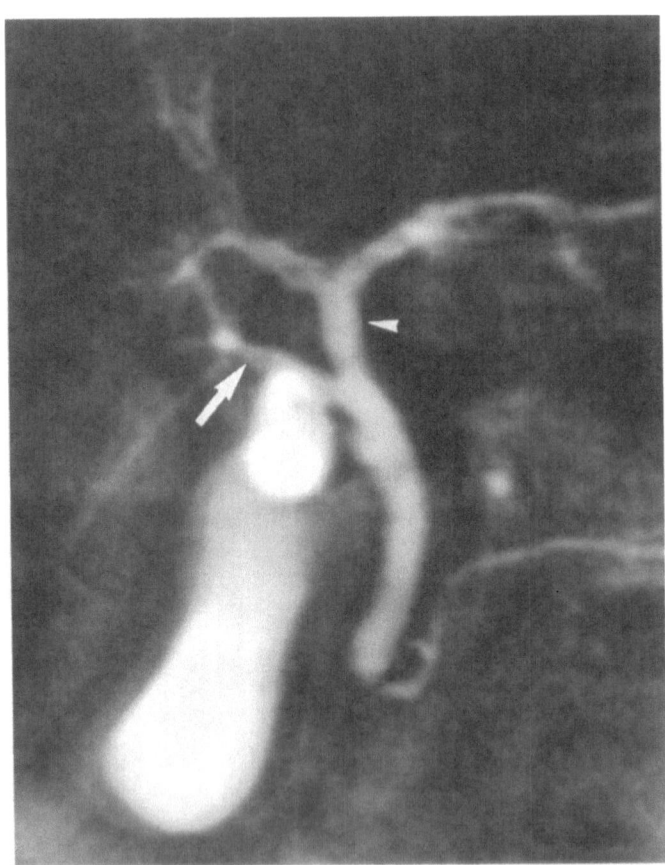

Fig. 3-3   Low cystic duct insertion

Single-slice MRCP shows the low insertion of the cystic duct (*arrow*) in the common hepatic duct.

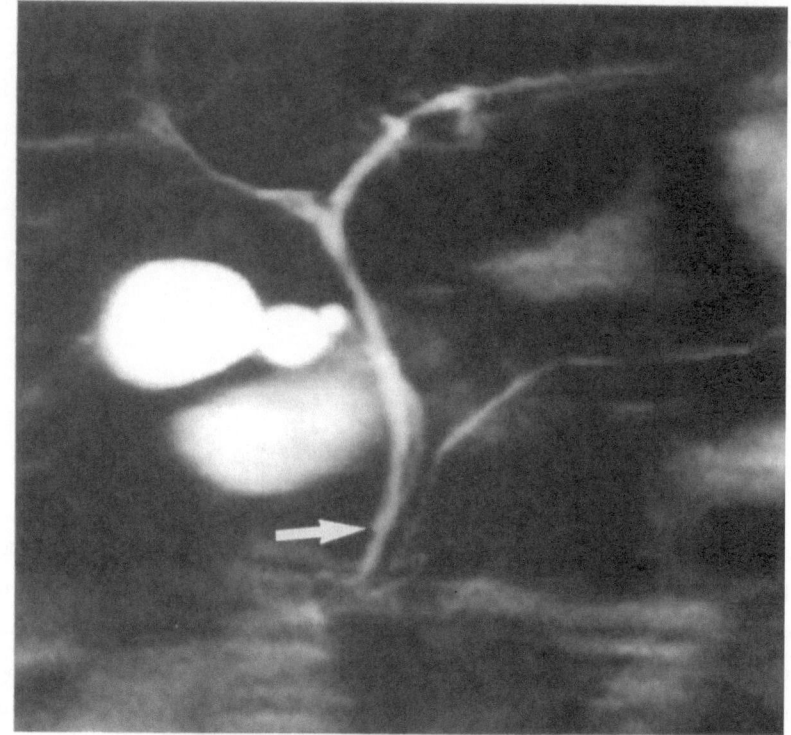

Fig. 3-4   Left-sided gallbladder

Single-slice MRCP shows the left-sided gallbladder (*arrow*) [10].

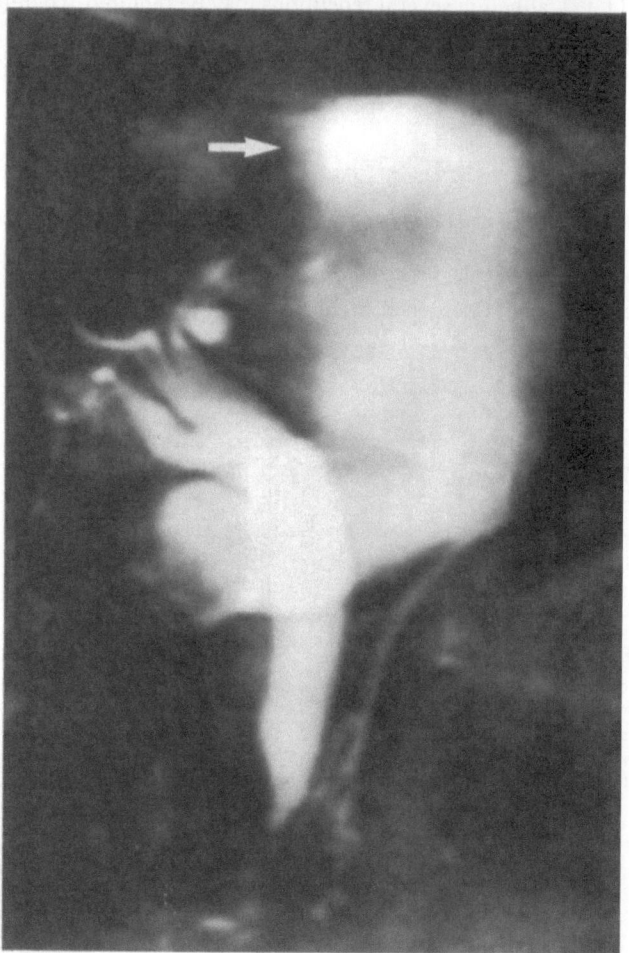

3.3 Pancreatobiliary Maljunction

3.3.1   Pancreatobiliary Junction

MRCP enables visualization of the morphology and contractile activity of the Vaterian sphincter complex [11]. In our study of 177 patients with suspected pancreatobiliary diseases, the success rate of imaging the pancreatobiliary junction by MRCP was 96% [12]. Since the scanning time for the single-shot fast spin echo technique of a single thick slice (2-D acquisition) is as short as 2 to 4 s, images of the relaxation and contraction phases of the sphincter of Oddi can be obtained almost in real time by dynamic MRCP study [11,12], although several repetitions may be required. Therefore, MRCP can be used to evaluate whether the action of the sphincter muscle (Oddi) functionally affects the union of the pancreatic and biliary ducts (Fig. 3-5 a, b).

Fig. 3-5   Single-slice MRCP shows pancreatobiliary junction (dynamic study)

a. Sphincter of Oddi relaxed (*arrow*).
b. Sphincter of Oddi contracted (*arrow*).

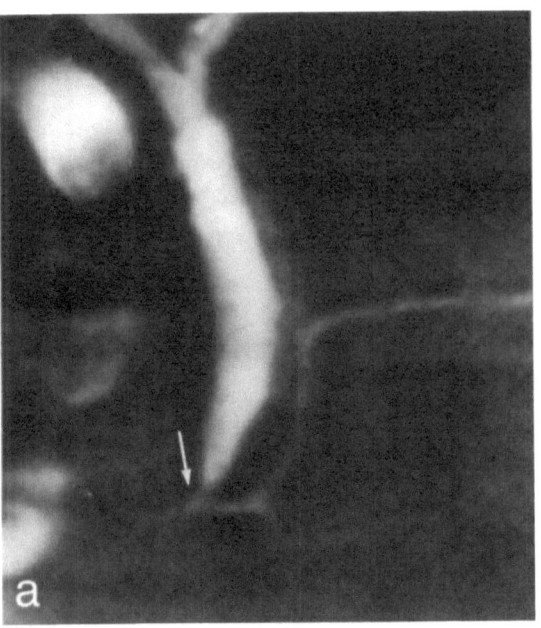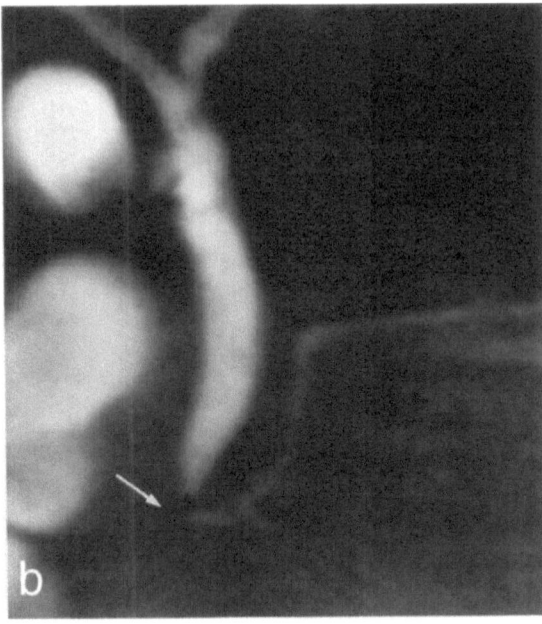

3.3.2   Pancreatobiliary Maljunction

Pancreatobiliary maljunction is a congenital anomaly of the arrangement of the pancreatobiliary ductal system, defined as union of the pancreatic and the biliary ducts outside the duodenal wall, forming a long common channel 15 mm or greater on an ERCP image [13-15]. Pancreatobiliary maljunction is present in 90%-100% of patients with congenital choledochal cyst [14,16]. In patients with pancreatobiliary maljunction, pancreatic juice refluxes into the common bile duct or bile juice regurgitates into the pancreatic duct, because the action of the sphincter muscle (Oddi) does not functionally affect the union. As a result, various pathological conditions occur in the biliary tract and the pancreas. Cholangitis, bile duct dilatation, cholelithiasis, biliary cancer, pancreatitis, and pancreatolithiasis are some of the known complications of pancreatobiliary maljunction [14,16,17], and the incidence of biliary carcinoma is reported to be 15.6%-36.0% in adult patients [16,17]. Therefore early diagnosis and surgical treatment are important to prevent development of carcinomas of the biliary tract.

MRCP is a reliable imaging method for diagnosing pancreatobiliary maljunction [11,18-21]. When a 15-mm common channel length is a diagnostic criterion for pancreatobiliary maljunction on MRCP, the sensitivity and specificity has been reported to be 82% and 100%, respectively [21]. In our study of 177 patients with suspected pancreatobiliary diseases, the sensitivity and specificity in diagnosing pancreatobiliary maljunction were 100% and 98%, respectively [11]. We also demonstrated with dynamic MRCP that the sphincter action only reaches below the anomalous pancreatobiliary junction [11]. Therefore, pancreatobiliary maljunction may be diagnosed on MRCP by a finding that the sphincter of Oddi has no effect on the pancreatobiliary junction.

Fig. 3-6   Pancreatobiliary maljunction without choledochal cyst

a. Single-slice MRCP shows abnormal pancreatobiliary junction (*arrow*) without cholangiectasis.
b. ERCP: Same finding as in MRCP (*arrow*).

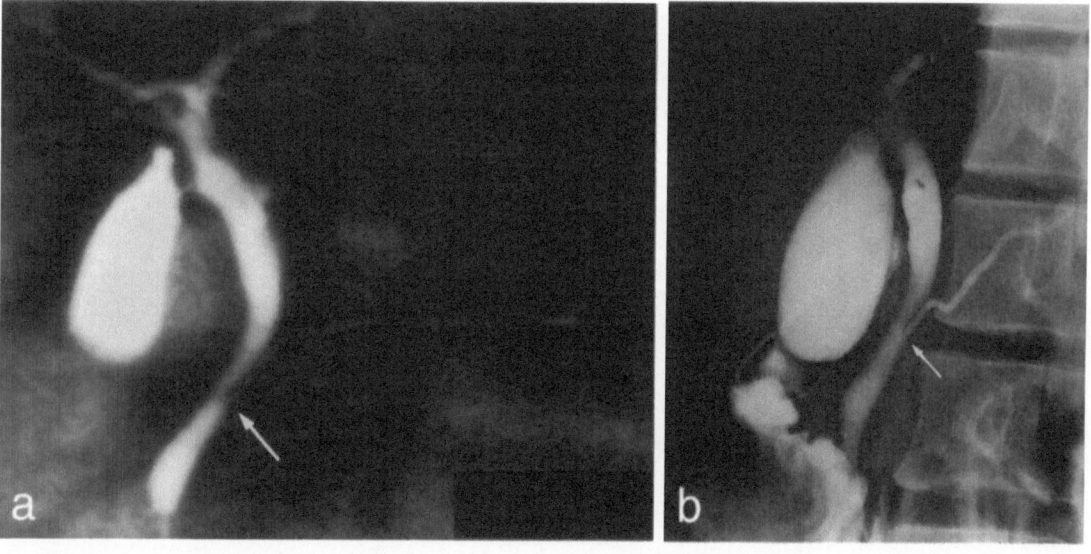

For the purpose of early diagnosis, MRCP should be performed in patients with pancreatitis of unknown cause or ultrasonographic findings indicating pancreatobiliary maljunction, such as bile duct dilatation, polypoid lesions of the gallbladder, or diffuse thickening of the gallbladder mucosa [14,16,17].

Fig. 3-7   Pancreatobiliary maljunction with choledocal cyst and early carcinoma of the gallbladder

a. Single-slice MRCP shows abnormal pancreatobiliary junction (*arrow*) with cholangiectasis.

b. Histology shows tubular adenocarcinoma of the gallbladder confined to the mucosa (early carcinoma of gallbladder).

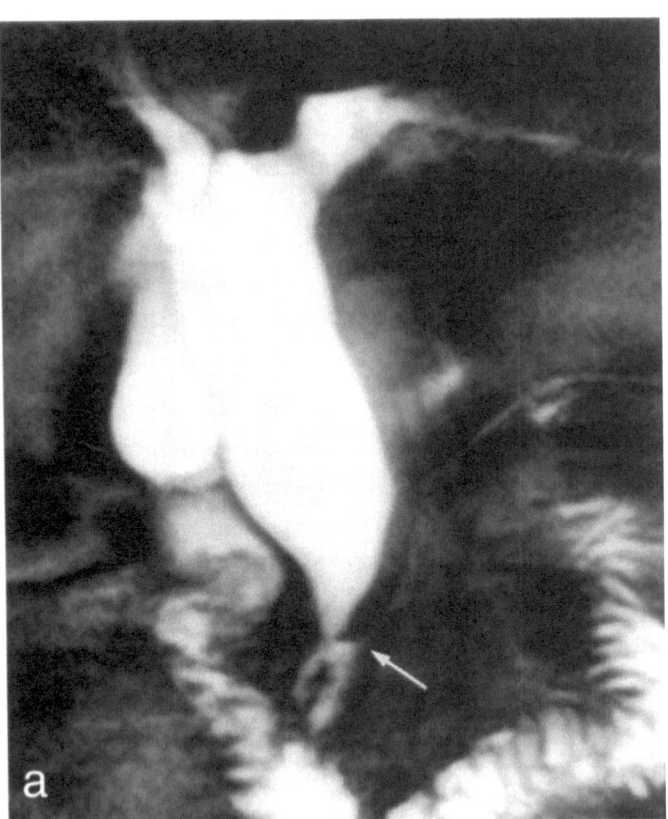

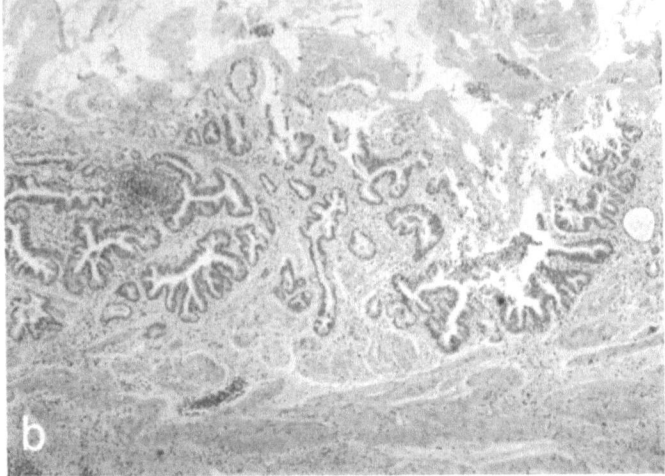

3.4 Cholelithiasis

In MRCP images, calculi are visualized as low-intensity areas (signal voids or filling defects) surrounded by hyperintense bile (Fig. 3-8b), although this appearance is not specific for calculi but may also represent tumors, blood clots, protein plugs, air bubbles (pneumobilia), or artifacts (e.g., flow void, surgical clip, etc.) (see section 3.11) [22,23]. Differentiation from pneumobilia is usually possible, because biliary air is located in nondependent portions of the biliary tract, whereas calculi tend to be located in dependent portions [24,25]. Some calculi contain a water component with long T2 relaxation[26,27] and demonstrate internal hyperintense areas [26,28].

With an MIP image, it is important to review the source image to confirm the presence of calculi (Fig. 3-8d), because signal voids produced by small stones may be obscured by the surrounding hyperintense bile (Fig. 3-8c). On the other hand, MRCP scanning with a single thick slice may fail to depict the stones in the gallbladder because of partial volume effects (Fig. 3-8b).

3.4.1   Cholecystolithiasis

Gallstones are shown as signal voids in the dependent part of the gallbladder. Calculi in the gallbladder are correctly diagnosed in 80% of cases using single thick slice MRCP [22], although accurate diagnosis of cholecystolithiasis is achieved by ultrasonography and the role of MRCP is therefore limited.

Because of the partial volume effects, cross-sectional images with multislice acquisition are more sensitive than projectional images with single thick slice acquisition in the diagnosis of cholecystolithiasis. In our experience of 368 MRCP studies, calculi in the gallbladder were correctly diagnosed in 71% with projectional images (using half-Fourier RARE with a single thick slice), and 89% with cross-sectional images (using half-Fourier RARE with multiple thin slices) [9].

MRCP may fail to detect stones in the contracted gallbladder or gallbladder with hyperconcentrated bile with accelerated T2 relaxation [22]. In such cases, additional T2-weighted tomograms covering the whole gallbladder may increase the diagnostic yield.

MRCP depicts the cystic duct in the majority of patients, and diagnosis of cystic duct calculi is feasible with an accuracy of 97% [29] (Fig. 3-10). However, the confirmation of patency of the cystic duct is not possible with MRCP.

Fig. 3-8   Cholecystocholedocholithiasis

a. ERCP shows filling defects in the common bile duct and gallbladder (*arrows*).

b. Single-slice MRCP shows stones as signal void areas in the common bile duct and the gallbladder (*arrows*).
However, not all the stones in the gallbladder are visible owing to the partial volume effect.

c. On an MIP-MRCP image, stones in the common bile duct and gallbladder are obscured by the surrounding
hyperintense bile.

d. By reviewing the source images of MIP, stones are found in the gallbladder and bile duct (*arrows*).

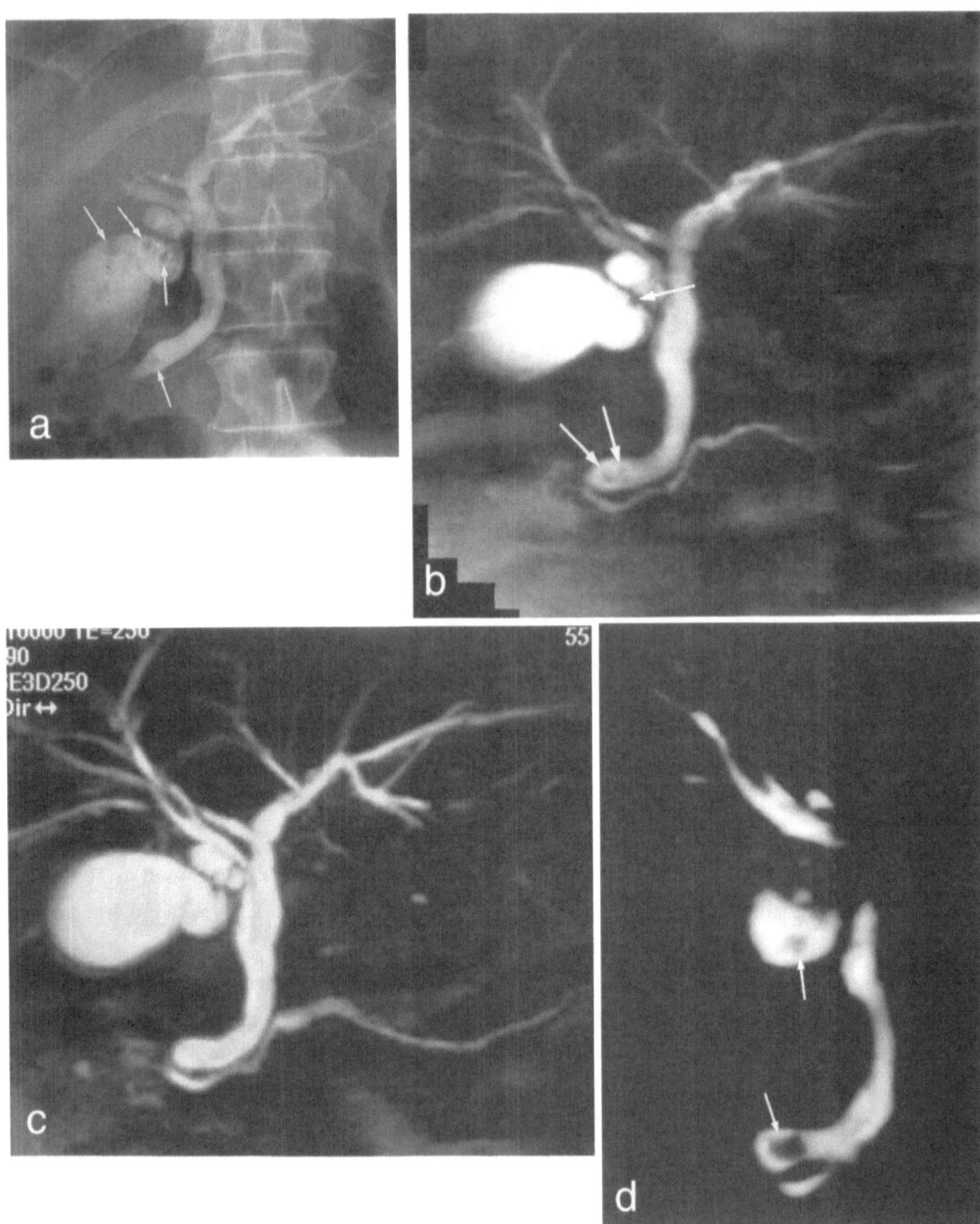

Fig. 3-9   Cholecystolithiasis

Single-slice MRCP shows multiple low-intensity areas in the gallbladder.

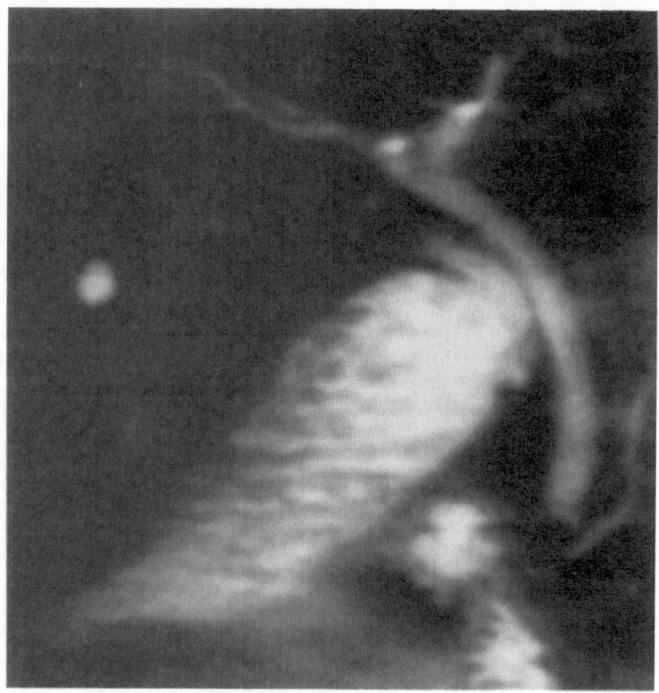

Fig. 3-10   Cystic duct calculi

a. Single-slice MRCP shows low intensity areas in the cystic duct (*arrow*) and inside the shrunken gallbladder (*open arrow*). Small stones are also seen in the distal portion of the common bile duct (*arrowhead*).

b. ERCP shows cystic duct obstruction due to the cystic duct stone (*arrow*) (after removal of the choledocholithiasis).

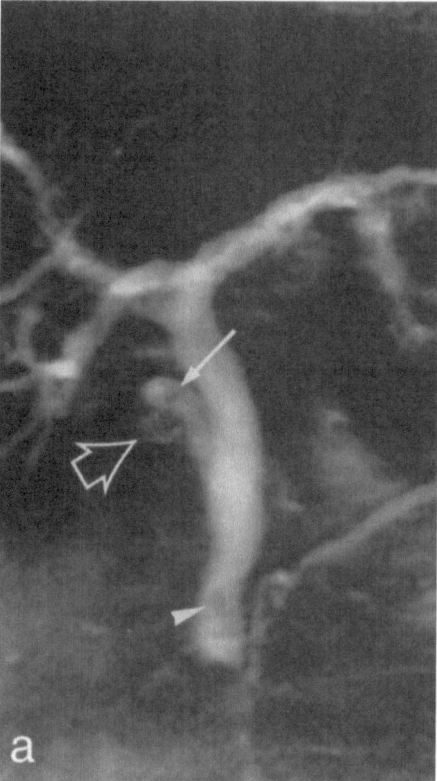
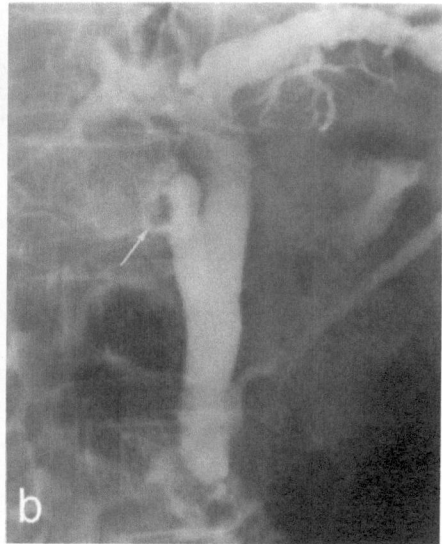

3.4.2   Choledocholithiasis

The sensitivity of MRCP in the diagnosis of choledocholithiasis has recently been reported to be 90%–100% [3,23,28,30]. Choledocholithiasis appears as low-intensity areas in the common bile duct. MRCP can depict stones at a minimum size of approximately 2–3 mm in diameter [3,24]. Pitfalls in the diagnosis of choledocholithiasis include pneumobilia (Fig. 3-42), hemobilia, protein plugs, polypoid tumor, surgical clips (Fig. 3-47), and flow effects (Fig. 3-46), all of which may create signal voids in the common bile duct. Axial MR images often allow the differentiation of pneumobilia, which floats anterior to bile, from choledocholithiasis, which lies in the dependent portion of the bile duct lumen [23].

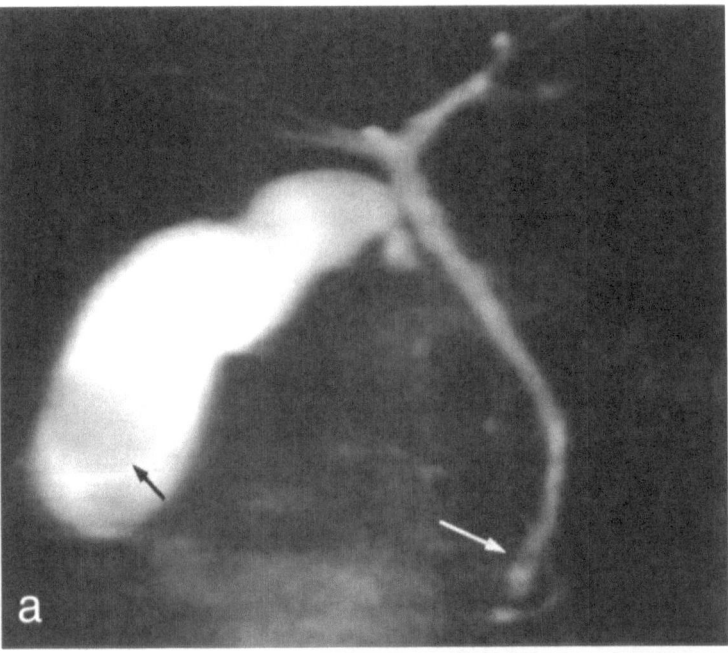

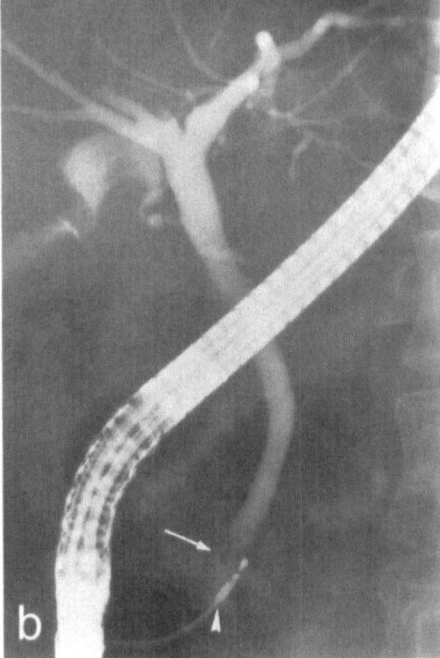

Fig. 3-11 Cholecystocholedocholithiasis (3 mm diameter)

a. Single-slice MRCP shows a small low-intensity area, consistent with a stone, in the distal common bile duct (*arrow*). The gallbladder also contains stones (*arrowhead*).
b. ERCP depicts a small stone, 3 mm in diameter (*arrow*), in the distal common bile duct, which was confirmed by intraductal ultrasonography (*arrowhead*).

Fig. 3-12 Cholecystocholedocholithiasis

Single-slice MRCP shows calculi in the distal common bile duct (*arrows*) and in the gallbladder. A dilated cystic duct indicates that gallbladder stones migrated into the common bile duct [31].

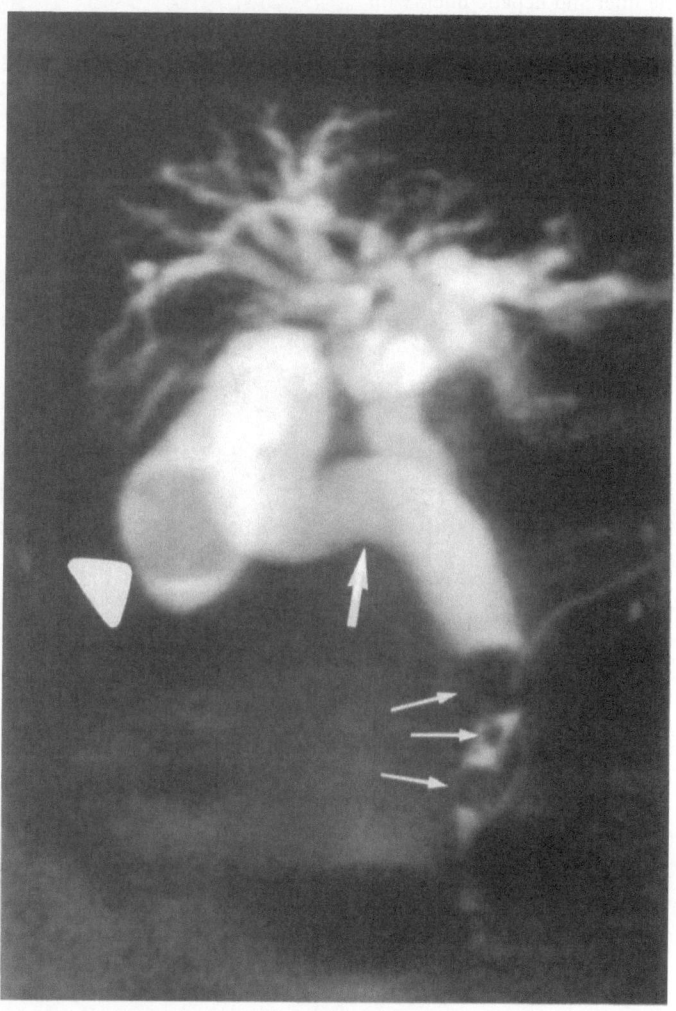

Fig. 3-13 Choledocholithiasis

a. Single-slice MRCP shows marked dilatation of the common bile duct and hepatic duct, with multiple stones (*arrows*).

b. ERCP shows choledocholithiasis (*arrows*). However, visualization of the hepatic duct is insufficient due to multiple stones.

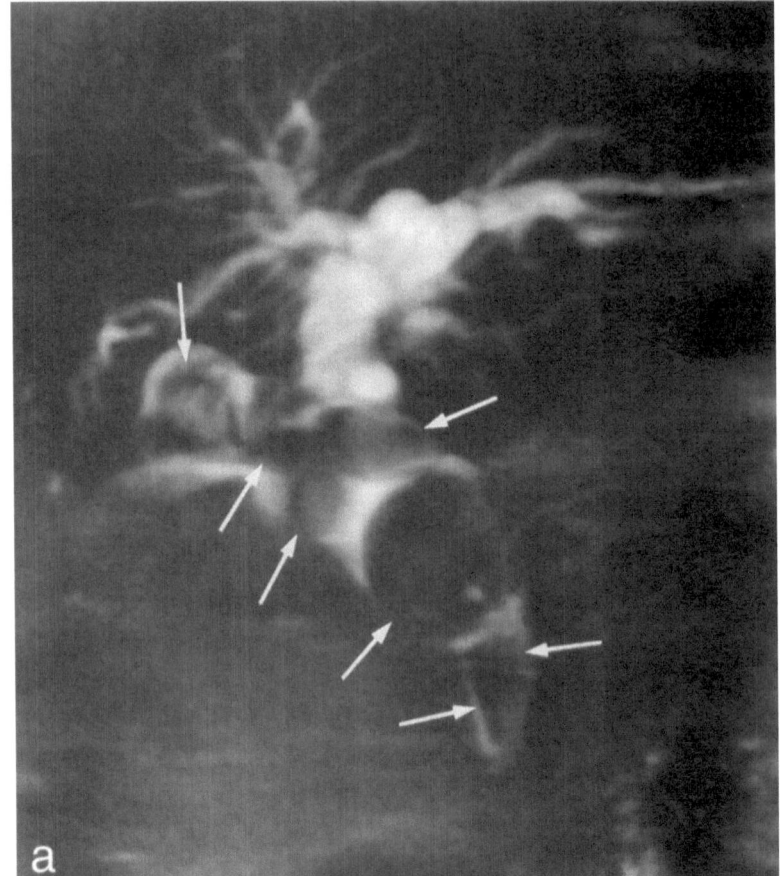

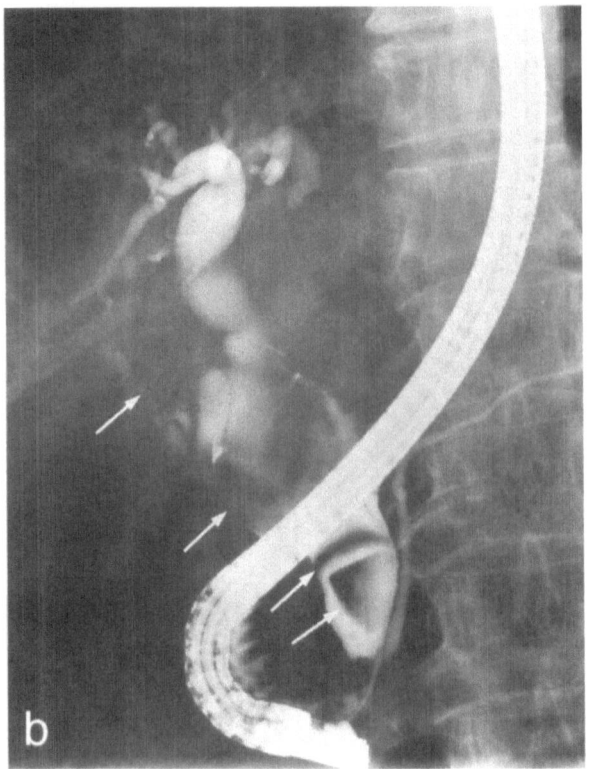

3.4.3   Intrahepatic Gallstones

The incidence of intrahepatic gallstones is 2.24% of cholelithiasis cases in Japan [32]. This condition is frequently complicated by stenosis or dilatation of the intrahepatic bile ducts. Moreover, intrahepatic cholangiocarcinomas are associated with intrahepatic gallstones in 2.4%–5.1% of cases [33, 34]. Therefore, detailed depiction of the intrahepatic bile ducts by direct cholangiography is mandatory for the management of intrahepatic gallstones, although US is effective for the diagnosis of intrahepatic gallstones.

In MRCP, intrahepatic gallstones are visualized as intraductal foci of low signal intensity [35]. However, MRCP may fail to demonstrate stones in the peripheral bile ducts or hepatic parenchyma, which are not surrounded by hyperintense bile. A pitfall in diagnosing intrahepatic gallstones with MRCP is pneumobilia, which is often seen in patients after biliary-enteric anastomosis.

MRCP provides detailed information about the morphological changes of the intrahepatic bile duct, including stenosis or dilatation, which is necessary for planning surgical treatment. MRCP also visualizes the intrahepatic bile ducts proximal to the site of the stone or stenosis, which cannot be imaged by direct cholangiography.

Fig. 3-14   Intrahepatic gallstones

a. Single-slice MRCP shows low intensity areas, consistent with stones, in the dilated intrahepatic bile duct (*arrows*).
b. ERCP confirms the presence of stones (*arrows*) in the dilated intrahepatic bile duct. However, the portion of the intrahepatic bile duct proximal to the obstruction is not imaged (*arrowhead*).

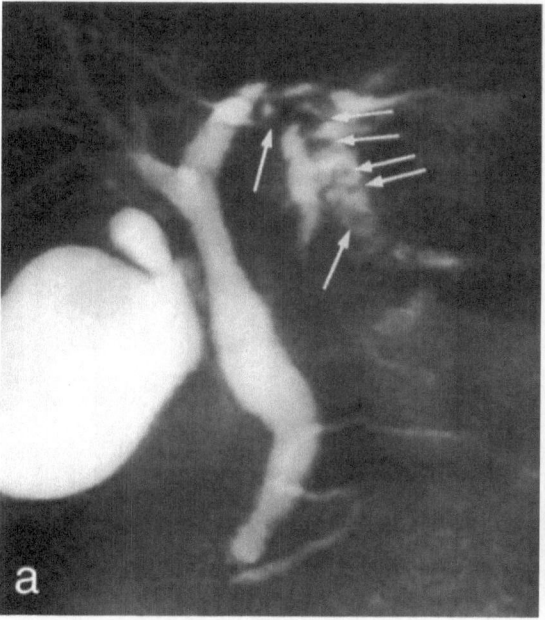
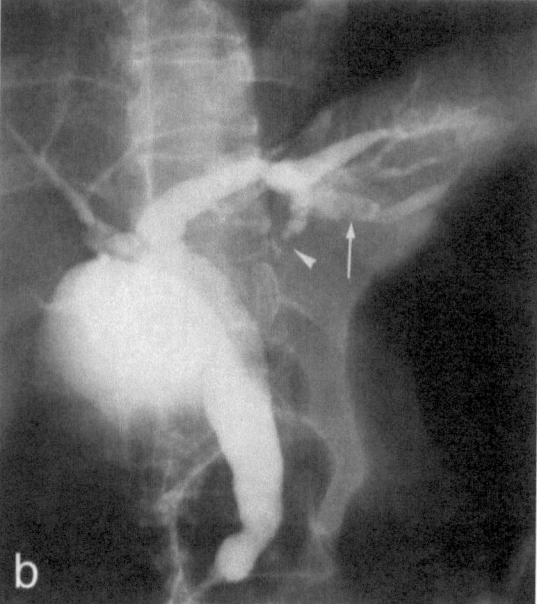

Fig. 3-15 Intrahepatic gallstones (recurrence after left lobectomy and papilloplasty)

a. Single-slice MRCP shows stenosis (*arrow*) in the right hepatic duct, and stones (*arrowheads*) in the dilated intrahepatic bile ducts. The common bile duct is filled with air.

b. ERCP confirms intrahepatic stones (*arrowheads*) and air in the common bile duct. The site of stenosis and its proximal portion of the intrahepatic bile duct are more clearly depicted with MRCP.

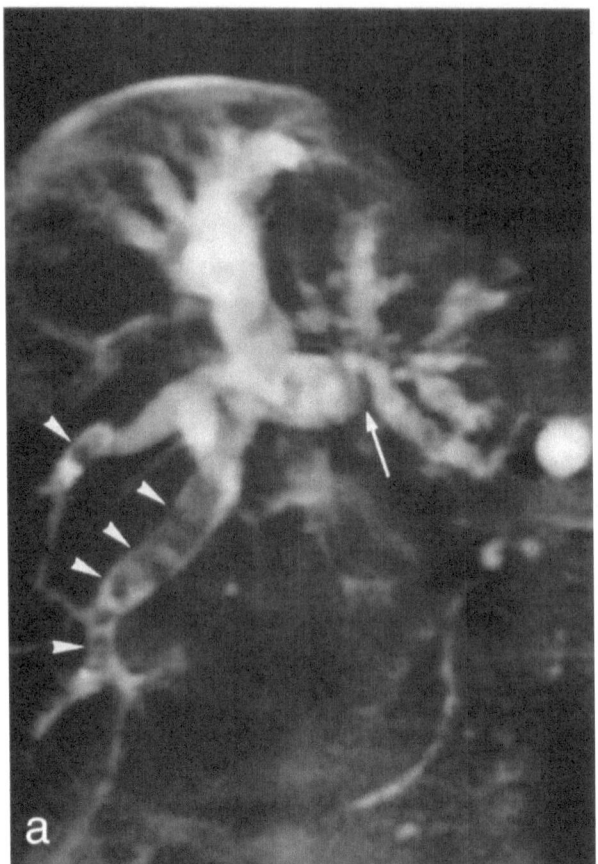

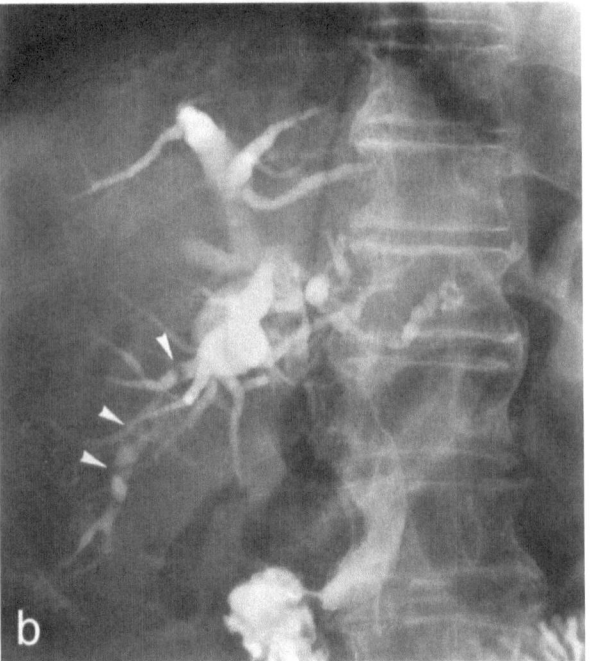

3.5 Acute Cholecystitis

Most cases of acute cholecystitis are caused by cystic duct obstruction due to stones in the neck of the gallbladder or in the cystic duct [36,37].

In the diagnosis of acute cholecystitis, MRCP is superior to ultrasonography (US) in the depiction of cystic duct calculi and calculi in the gallbladder neck, although US is superior to MRCP in evaluating gallbladder wall thickening [29]. Because MRCP detects water signals with high sensitivity, it can clearly depict inflammatory effusion associated with acute cholecystitis [38,39], although this sign is unreliable in patients with hepatitis or ascites where free fluid may be present around the gallbladder without intrinsic gallbladder disease [39]. Also, MRCP is not reliable in the diagnosis of gallbladder empyema [22].

In our experience, MRCP provides detailed information about acute cholecystitis; gallbladder distention, edema of the gallbladder wall, calculi in the gallbladder neck, cystic duct calculi, sludge in the gallbladder, inflammatory effusion, and pericholecystic abscess [38]. Nevertheless US is routinely used for the diagnosis of acute cholecystitis [40,41]. Therefore, MRCP supplements US and CT by providing additional information about acute cholecystitis.

Fig. 3-16 Acute cholecystitis

a. Ultrasonography shows stones (*arrow*) in the neck of gallbladder and gallbladder wall thickening (*arrowheads*).
b. Single-slice MRCP shows cystic duct obstruction due to calculi (*arrow*) in the gallbladder neck. Associated inflammatory effusion is present around the gallbladder (*arrowheads*). The distribution of the effusion is more clearly depicted than in US.

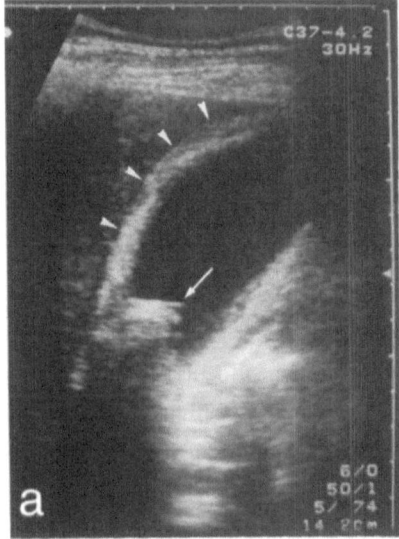
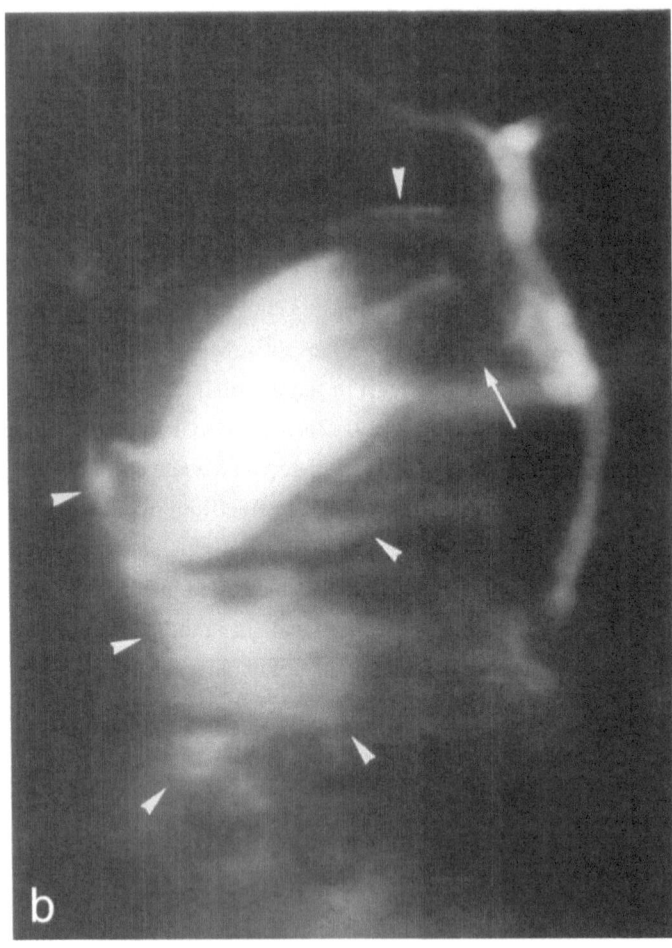

Fig. 3-17 Acute cholecystitis

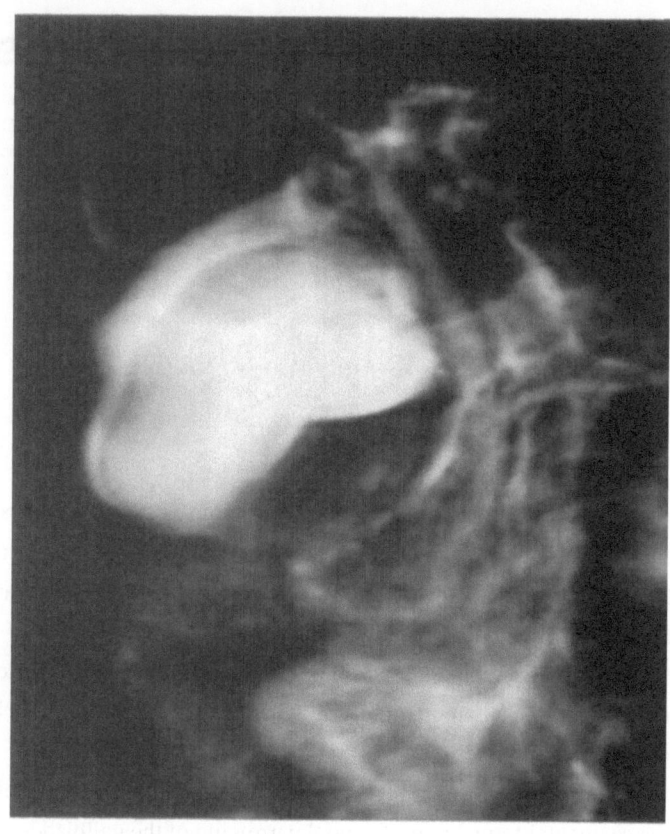

Fig. 3-18 Schema of Fig. 3-17

Single-slice MRCP shows gallbladder distention (*arrow*) with cystic duct obstruction. Inflammatory effusion is retained in the gallbladder bed (*arrowhead*), suggesting pericholecystic abscess. Effusion associated with cholecystitis spreads from the lesser omentum to the periportal region (*open arrowheads*), showing relatively widespread inflammation.

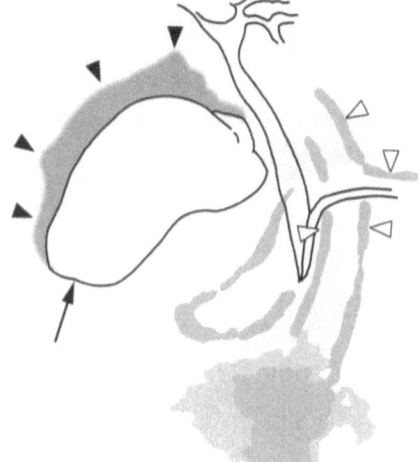

3.6 Adenomyomatosis of the Gallbladder

Adenomyomatosis (ADM) of the gallbladder is a nonneoplastic condition with mucosal hyperplasia and enlargement of the Rokitansky–Aschoff sinuses (RAS), always in association with proliferation of the smooth muscle. The condition may be diffuse, segmental, and localized (Fig. 3-19) [42-44]. ADM localized to the fundus is sometimes called adenomyoma [45]. The segmental form of ADM is reported to be associated with gallbladder cancer in 6.4% of cases [46].

In MRCP, demonstration of the RAS is essential in the diagnosis of ADM, as in conventional cholangiography. MRCP demonstrates the RAS as diverticular outpouchings of high signal intensity spot in the gallbladder wall [47,48].

In the diffuse form of ADM, the high-signal intensity spots of RAS surround the whole gallbladder. In the segmental form, circumferential narrowing of the gallbladder accompanied with RAS occurs, usually in the midportion of the gallbladder with "hour-glass" deformity [49]. The localized form often affects the fundus of the gallbladder and shows a sessile mass with the centrally umbilicated pit [50].

Fig. 3-19 Classification of adenomyomatosis of the gallbladder according to Jutras [42]

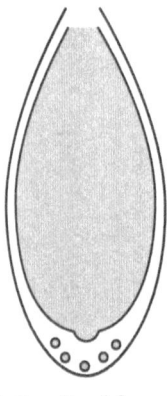

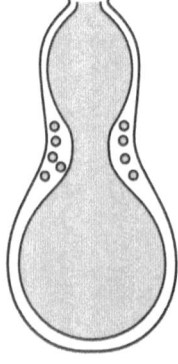

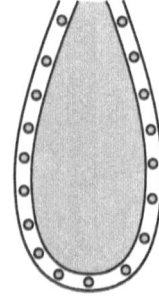

A: Localized form.   B: Segmental form.   C: Generalized form.

Fig. 3-20 Localized ADM

Single-slice MRCP shows a centrally umbilicated pit at the fundus of the gallbladder.

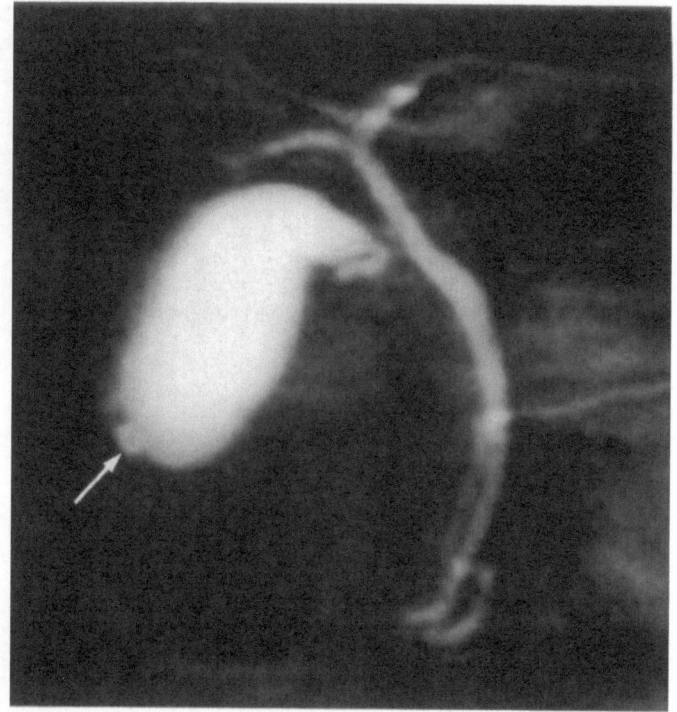

Fig. 3-21 Segmental ADM

Single-slice MRCP shows circumferential narrowing of the gallbladder accompanied by RAS in the midportion of the gallbladder with "hour-glass" deformity.

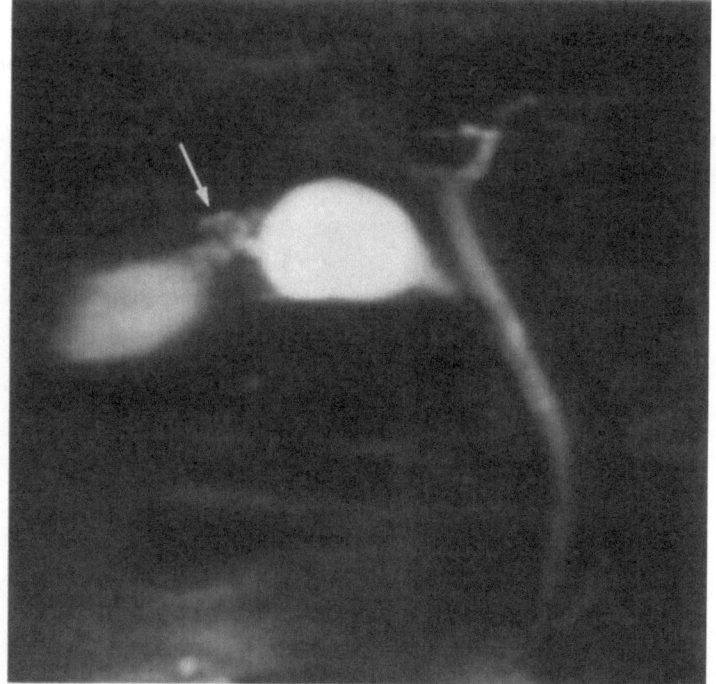

Fig. 3-22 Generalized ADM

Single-slice MRCP shows the RAS
surrounding the whole gallbladder.

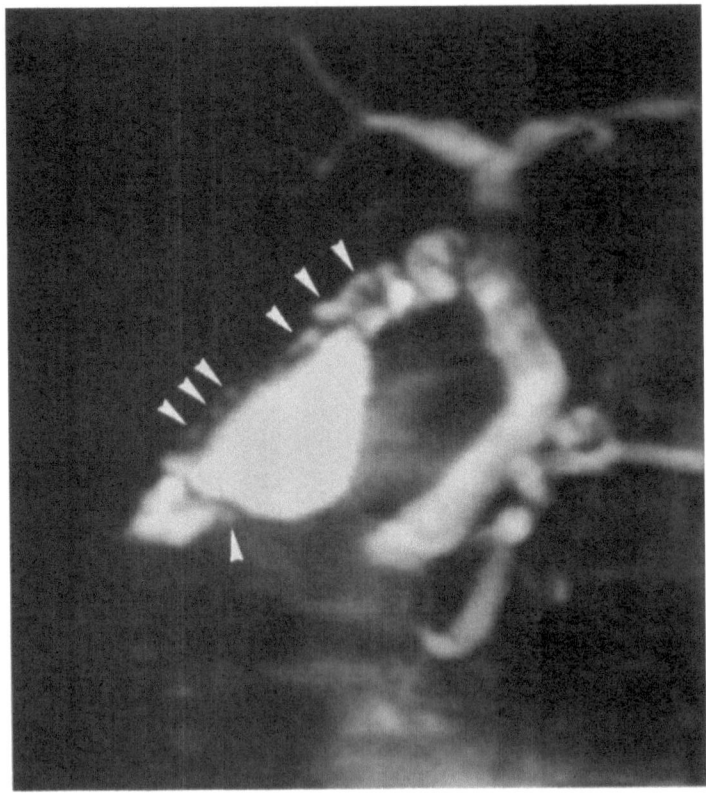

Fig, 3-23 Mixed type of ADM

a. Ultrasonography shows narrowing at the midportion of the gallbladder (*bold arrow*) and diffuse wall thickening
(*arrowheads*). A stone is seen at the neck (*arrow*).

b. Single-slice MRCP shows narrowing at the midportion of the gallbladder (*arrow*) with multiple RAS (*arrowheads*) from the midportion to the fundus of the gallbladder.

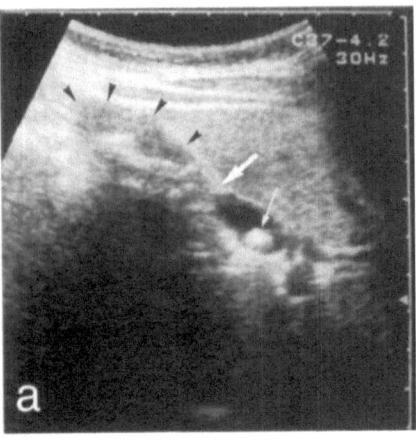

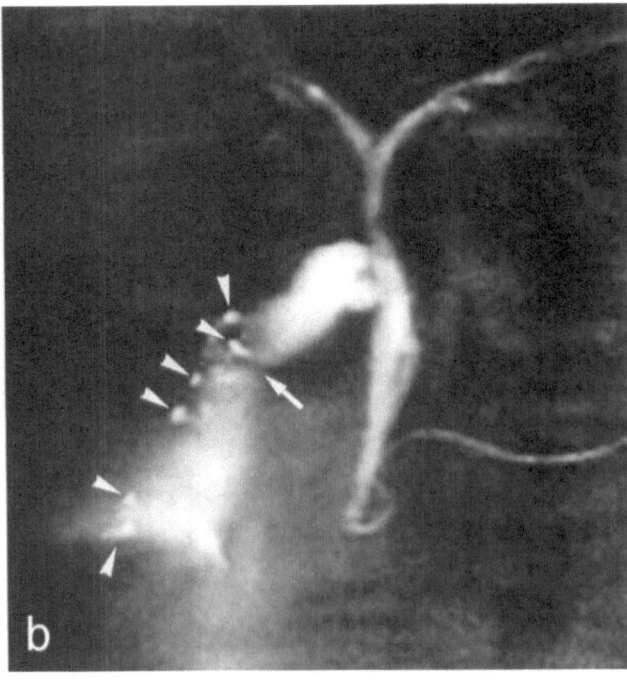

3.7 Polypoid Lesions of the Gallbladder

Polypoid lesions of the gallbladder include cholesterol polyps, hyperplastic polyps, adenomyomatosis, adenomas, and adenocarcinomas. Most of the early gallbladder carcinomas are presented as polypoid lesions, and 88% of those exceed 10 mm in diameter [51, 52].

In the diagnosis of polypoid lesions of the gallbladder, US is the most sensitive and reliable method. However, MRCP can depict polypoid lesions as low-intensity areas in the gallbladder.

Fig. 3-24 Early gallbladder carcinoma

a. Single-slice MRCP shows multiple and irregular low-intensity areas in the gallbladder (*arrow*) which exceed 10 mm in diameter.

b. Endoscopic ultrasonography (EUS) shows broad elevated lesions (*arrowheads*) with intact gallbladder wall.

c. Histology shows papillary adenocarcinoma limited to the mucosa.

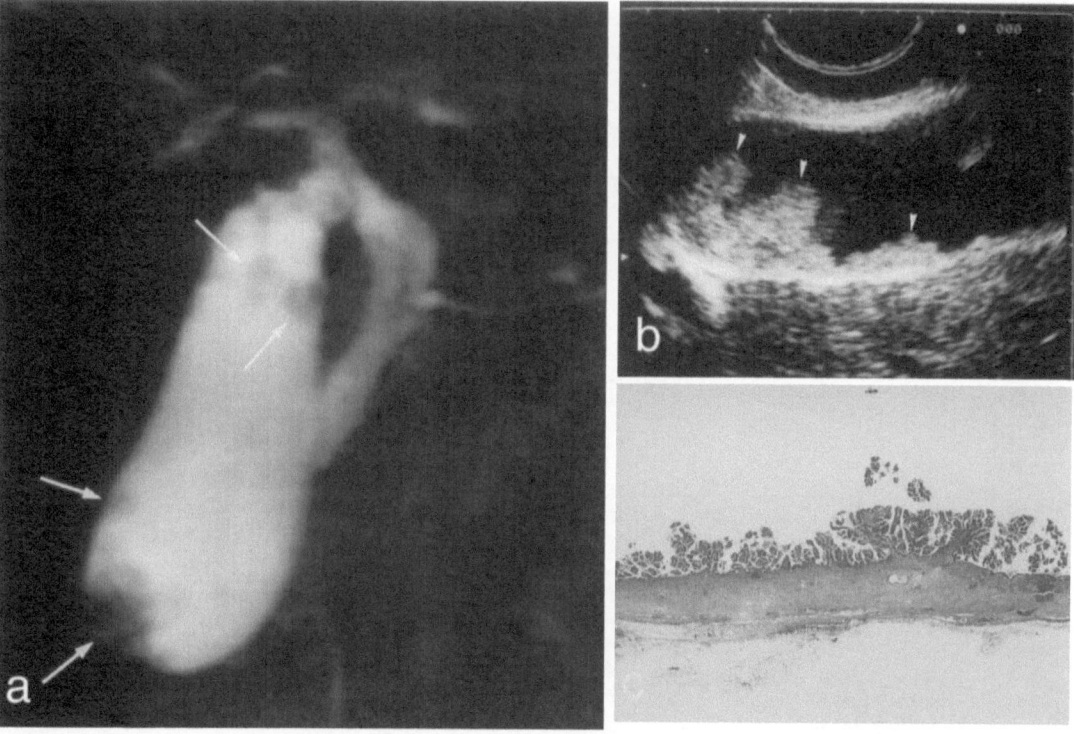

3.8 Carcinomas of the Biliary Tract

3.8.1   Carcinomas of the Gallbladder

In patients with gallbladder carcinoma, direct cholangiography may not be able to visualize the cystic duct and gallbladder due to tumor invasion. MRCP advantageously provides a projectional image of the gallbladder carcinoma as filling defects or irregular margin of the gallbladder wall.

Because diagnosis of gallbladder carcinoma is accurately established by US, endoscopic US, or CT, the role of MRCP is limited to preoperative evaluation of the site and extension of tumor invasion of the bile duct [53] (Fig. 3-25).

Furthermore, because carcinomas of the gallbladder are frequently associated with pancreatobiliary maljunction, MRCP may play an important role in the early diagnosis of gallbladder carcinoma [47] (Fig. 3-7).

Fig. 3-25 Gallbladder carcinoma with invasion of the common bile duct

a. CT demonstrates gallbladder carcinoma with invasion of the liver (*arrowhead*).
Intrahepatic bile ducts are not dilated, and bile duct invasion of the carcinoma is not assessed.
b. Single-slice MRCP shows stenosis of the common bile duct (*arrow*) by tumor invasion. Proximal bile ducts are
not dilated.

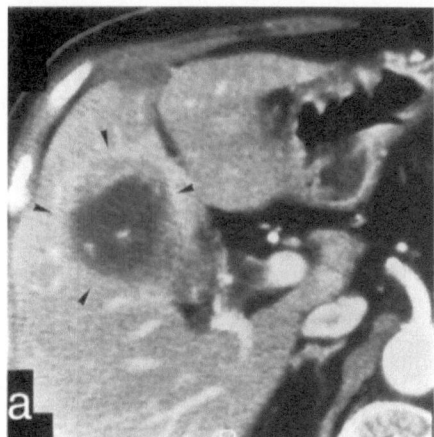
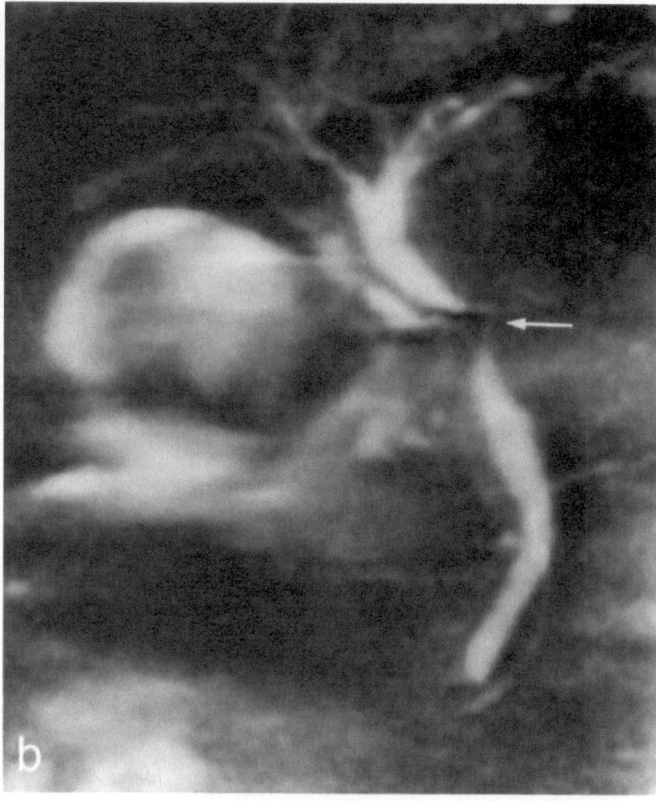

3.8.2   Extrahepatic Bile Duct Carcinomas

MRCP can diagnose the presence and the level of bile duct obstruction with an accuracy of 98% [3, 24]. Compared with ERCP and percutaneous transhepatic cholangiography, MRCP has the advantage of depicting the biliary tract both proximal and distal to the site of obstruction without the risk of sepsis that may occur by injecting the contrast material in the bile duct proximal to the obstruction [54]. Therefore, MRCP is useful to determine tumor extension along the biliary tract and the intrahepatic bile ducts, and makes it possible to identify tumor invasion in the various bile duct segments (see section 3.8.4)

MRCP depicts the biliary tract and pancreatic duct simultaneously, and is therefore useful for the differentiation of bile duct stenosis secondary to pancreatic diseases [55].

Fig. 3-26 Extrahepatic bile duct carcinoma

a. Three-dimensional MRCP with MIP shows a stenosis in the common bile duct (*arrow*) with proximal duct dilatation. Bile duct stenosis secondary to pancreatic diseases can be ruled out, because the pancreatic duct is normal.

b. Percutaneous transhepatic cholangiography shows an obstruction in the common bile duct (*arrow*), although the distal portion of the common bile duct is not opacified.

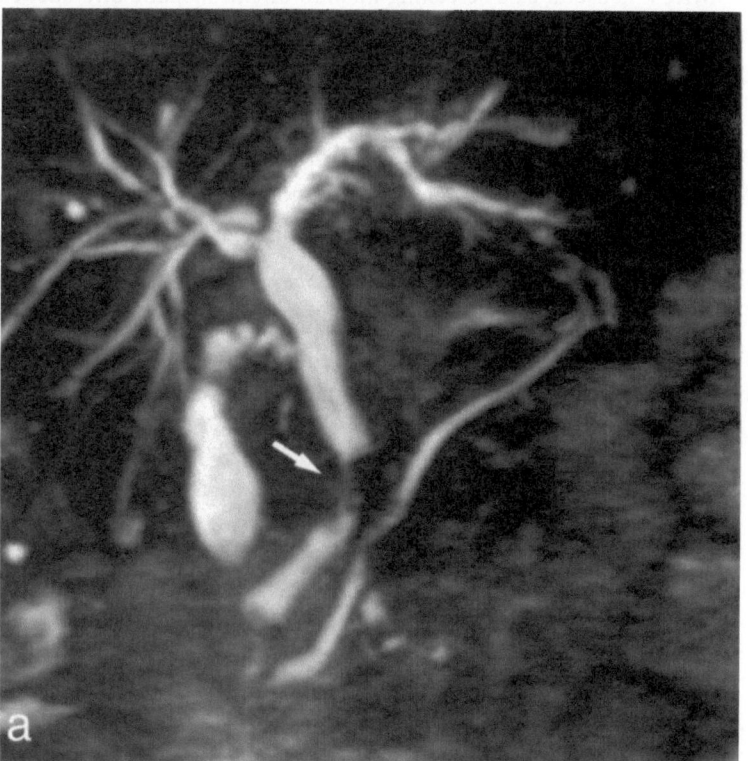

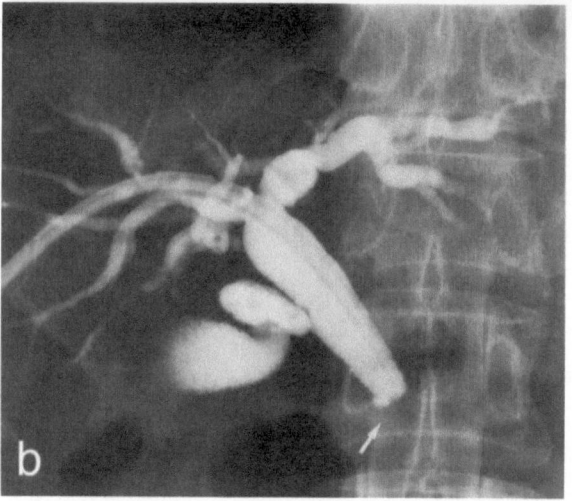

3.8.3   Early Carcinomas of the Extrahepatic Bile Duct

Early carcinomas of the extrahepatic bile duct are defined as tumors limited to the mucosa or the fibromuscular layer. Early carcinomas have a favorable postoperative prognosis with a 5-year survival rate of 80% [56].

MRCP is useful as a screening method of extrahepatic bile duct carcinomas at an early stage [57], because it can be conducted conveniently and safely in the outpatient clinic and has excellent accuracy in the diagnosis of bile duct stenosis and filling defects.

In our prospective study of 412 patients suspected of pancreatobiliary diseases, accuracy for bile duct stenosis and filling defects was 98% and 92%, respectively. Twelve patients were eventually diagnosed by MRCP to have extrahepatic bile duct carcinomas, including three patients without jaundice and two patients with carcinomas limited to the fibromuscular layer [3]. In these patients, the carcinomas were depicted as a short stricture or a filling defects of the common bile duct with slight proximal duct dilatation. However, differential diagnosis of cholangiocarcinoma from benign stricture or filling defects, such as postoperative bile duct stenosis (see section 3.9.5) or benign bile duct tumors, is difficult, and cytology or biopsy may be required for accurate diagnosis.

Fig. 3-27 Extrahepatic bile duct carcinomas at an early stage

a. Single-slice MRCP shows stenosis in the common bile duct (*arrow*) with slight proximal duct dilatation.

b. Histology shows an invasive tubular adenocarcinoma confined to the fibromuscular layer.

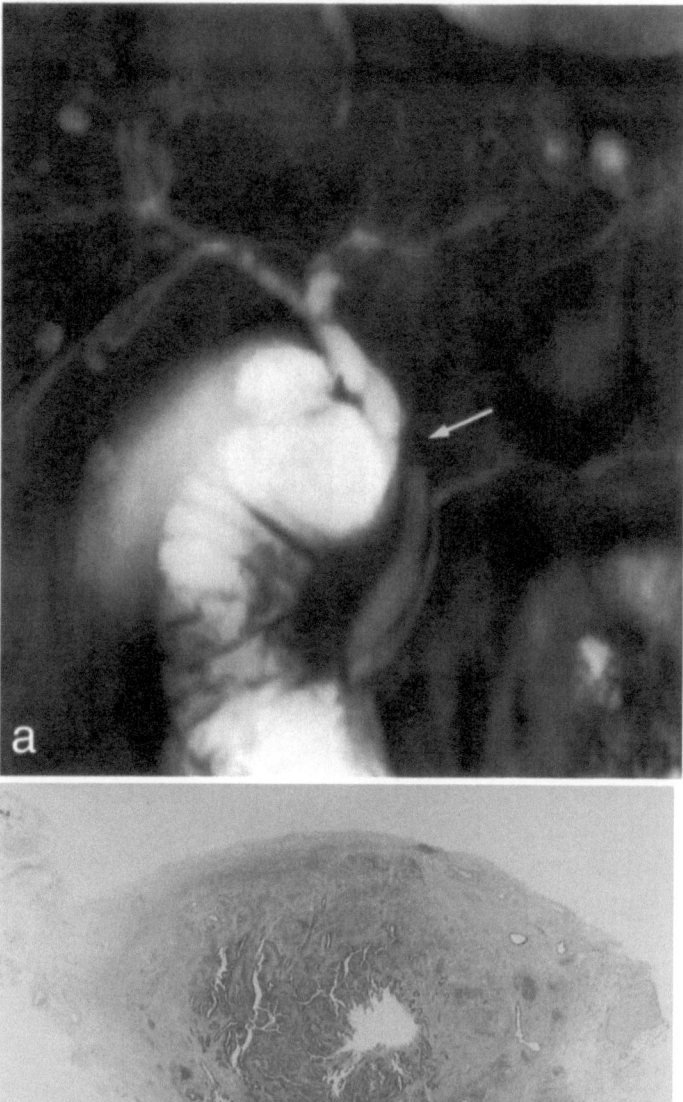

Fig. 3-28 Extrahepatic bile duct carcinomas at an early stage

a. Single-slice MRCP shows an irregular low-intensity area in the common bile duct (*arrow*) with slight proximal duct dilatation.

b. Coronal section image of contrast-enhanced T1-weighted gradient echo sequence shows an irregular elevated lesion in the common bile duct.

c. Histology confirms a papillary adenocarcinoma limited to the fibromuscular layer.

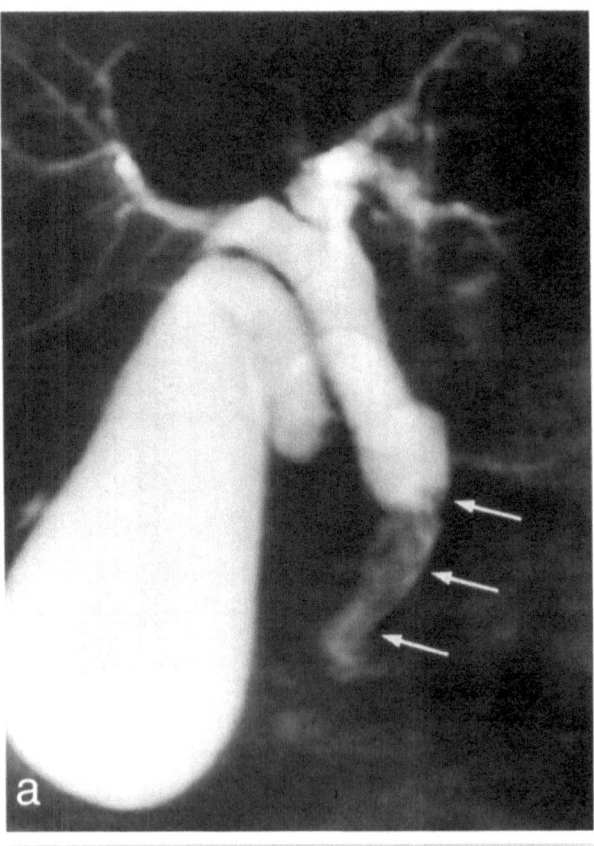

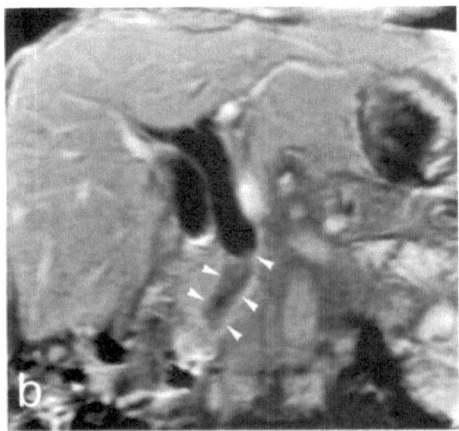

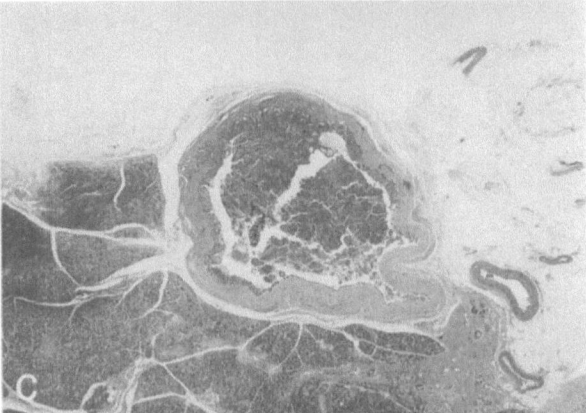

3.8.4   Hilar Cholangiocarcinomas (Klaskin's Tumors)

Hilar cholangiocarcinomas (Klaskin's tumors) were reported by Klaskin in 1965 as tumors that occur at the confluence of the right and left hepatic ducts. Recently, advanced surgical techniques and accurate preoperative tumor staging improved the survival rates [58].

MRCP demonstrates the tumor as a hilar stricture and extension along the intrahepatic and extrahepatic bile ducts, and can depict the isolated segments of the intrahepatic bile ducts that are not opacified with direct cholangiography. By rotating 3D-MIP images, identification of each branch of the intrahepatic bile duct is possible, and tumor invasion of the intrahepatic bile ducts including the caudate branches can be diagnosed accurately [9].Therefore, MRCP is useful to determine the resectability of hilar cholangiocarcinomas [59].

Fig. 3-29 Hilar cholangiocarcinoma

a. Three-dimensional MRCP with MIP shows the hilar stricture (*bold arrow*) with proximal dilatation of the left intrahepatic bile duct (*arrowhead*) as well as the right anterior segment of the intrahepatic bile duct (*arrow*).

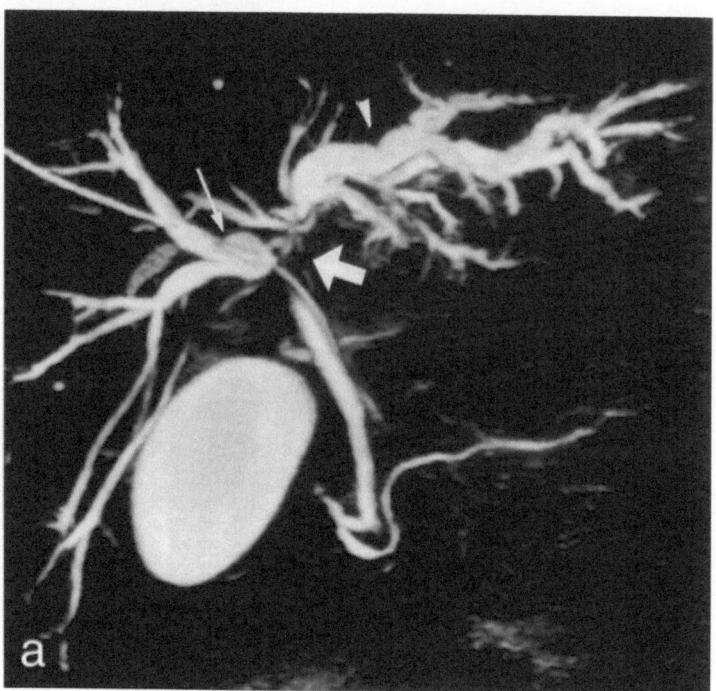

Fig. 3-29 *continued*

b. By rotating the MIP image, the caudal branches (*arrow*) can be evaluated.

c. Percutaneous transhepatic cholangiography shows stenosis in the hilar bile duct (*arrow*). However, intrahepatic bile ducts are not adequately visualized and proximal extension of the tumor is not apparent compared with MRCP.

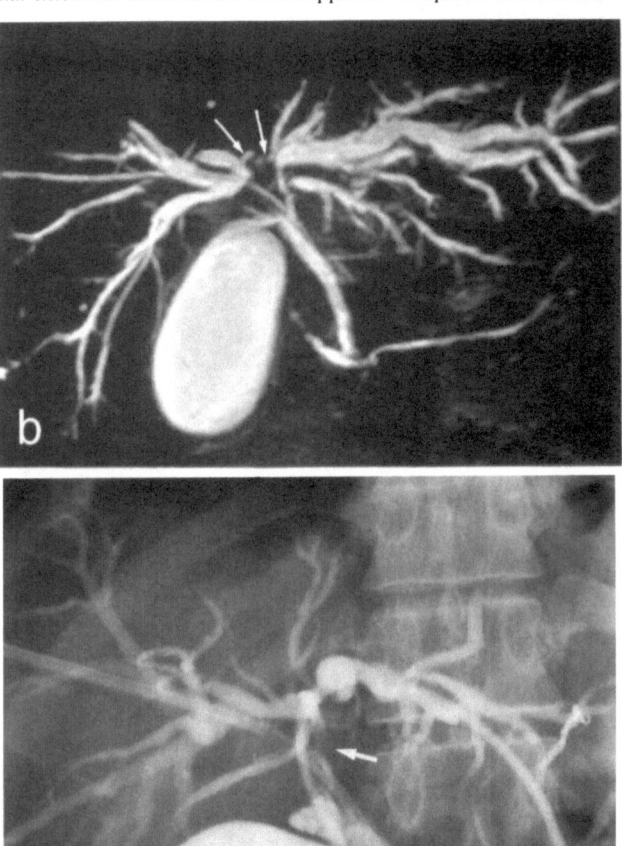

3.8.5   Carcinomas of the Papilla of Vater

MRCP reveals filling defects or stenosis of the distal common bile duct [3,60]. In general, bile duct dilatation is seen in 75% of cases, and pancreatic duct dilatation in 67% [61]. Because MRCP does not permit direct visualization and biopsy of the papilla of Vater, differential diagnosis of carcinoma of the papilla of Vater from impacted calculus in the distal common bile duct is sometimes difficult. Therefore, gastrointestinal endoscopy is the most accurate diagnostic method for carcinomas of the papilla of Vater. However, MRCP can depict tumor extension along the common bile duct and pancreatic duct noninvasively and is useful for evaluating the tumor resectability.

Fig. 3-30 Carcinoma of the papilla of Vater

a. Single-slice MRCP shows a filling defect in the distal common bile duct, pancreatic duct (*arrowheads*), and duodenum (*arrows*). Proximal duct dilatation is seen in both the biliary tract and pancreatic duct. The percutaneous transhepatic cholangiographic drainage (PTCD) tube is shown as a continuous high-intensity structure (*bold arrow*).

b. Percutaneous cholangiography demonstrates only filling defects (*arrow*) in the distal common bile duct.

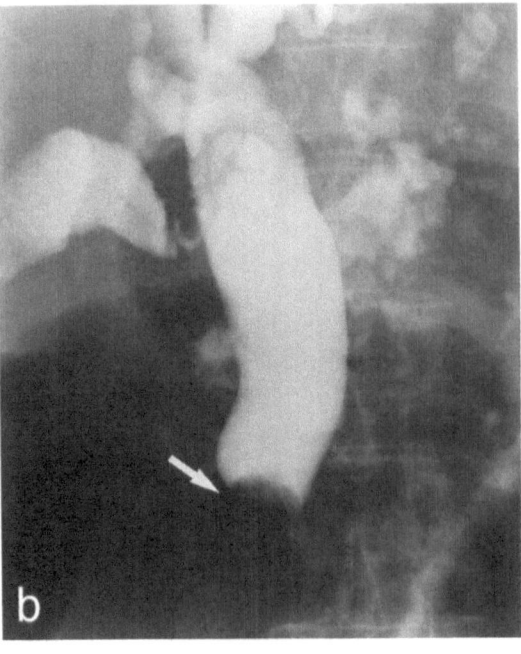

Fig. 3-31 Carcinoma of the papilla of Vater

Single-slice MRCP shows stenosis of both the distal common bile duct and pancreatic duct (*bold arrow*) with proximal duct dilatation. The PTCD tube appears as a continuous line (*arrow*).

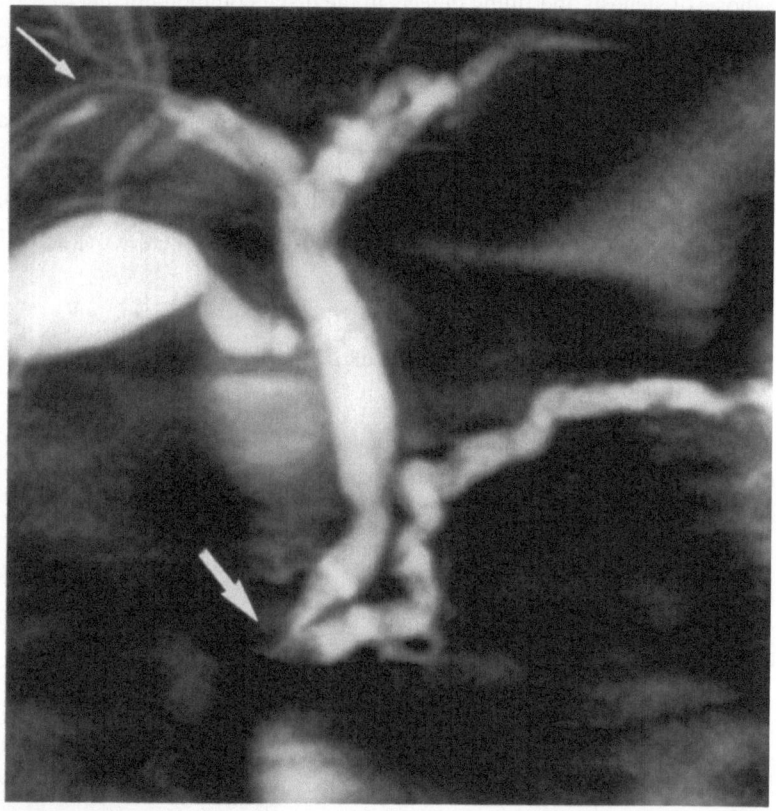

3.8.6   Early Carcinomas of the Papilla of Vater

MRCP is useful in the early diagnosis of carcinomas of the papilla of Vater because it is accurate in the diagnosis of stenosis and filling defects in the distal common bile duct.

In our prospective study of 412 patients suspected of pancreatobiliary diseases, accuracy rates for ampullary stenosis and filling defects was 98% and 92%, respectively. Six patients with carcinomas of the papilla of Vater were diagnosed by MRCP, including three patients without jaundice and two patients with carcinomas limited to the sphincter of Oddi [3]. A small polypoid lesion or mild stenosis of the pancreatobiliary ducts at the ampulla with slight proximal dilatation were seen on MRCP. Differential diagnosis of carcinoma from benign stenosis or dysfunction of the sphincter of Oddi was sometimes difficult, and the diagnosis had to be confirmed by endoscopy.

Fig. 3-32 Early carcinoma of the papilla of Vater

a. Single-slice MRCP shows filling defects in the distal common bile duct (*arrow*) with no proximal duct dilatation.
b. ERCP shows filling defects in the distal common bile duct (*arrow*) as in MRCP.
c. Histology shows carcinoma of the papilla of Vater confined to the mucosa.

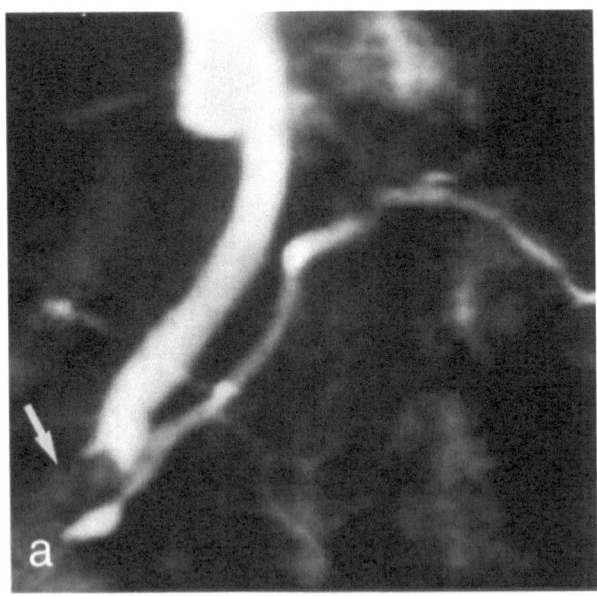

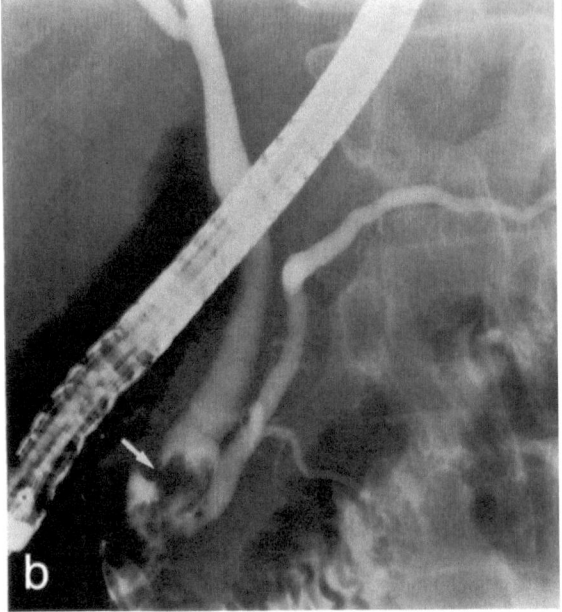

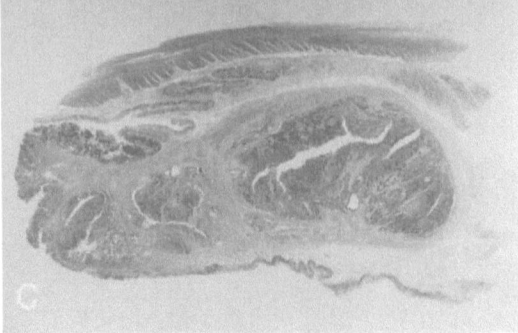

Fig. 3-33 Early carcinoma of the papilla of Vater

a. Single-slice MRCP shows stenosis of the ampullary bile duct (*arrow*) with mild proximal duct dilatation.

b. ERCP shows the same findings as in MRCP (*arrow*).

c. Histology shows carcinoma of the papilla of Vater confined to the mucosa.

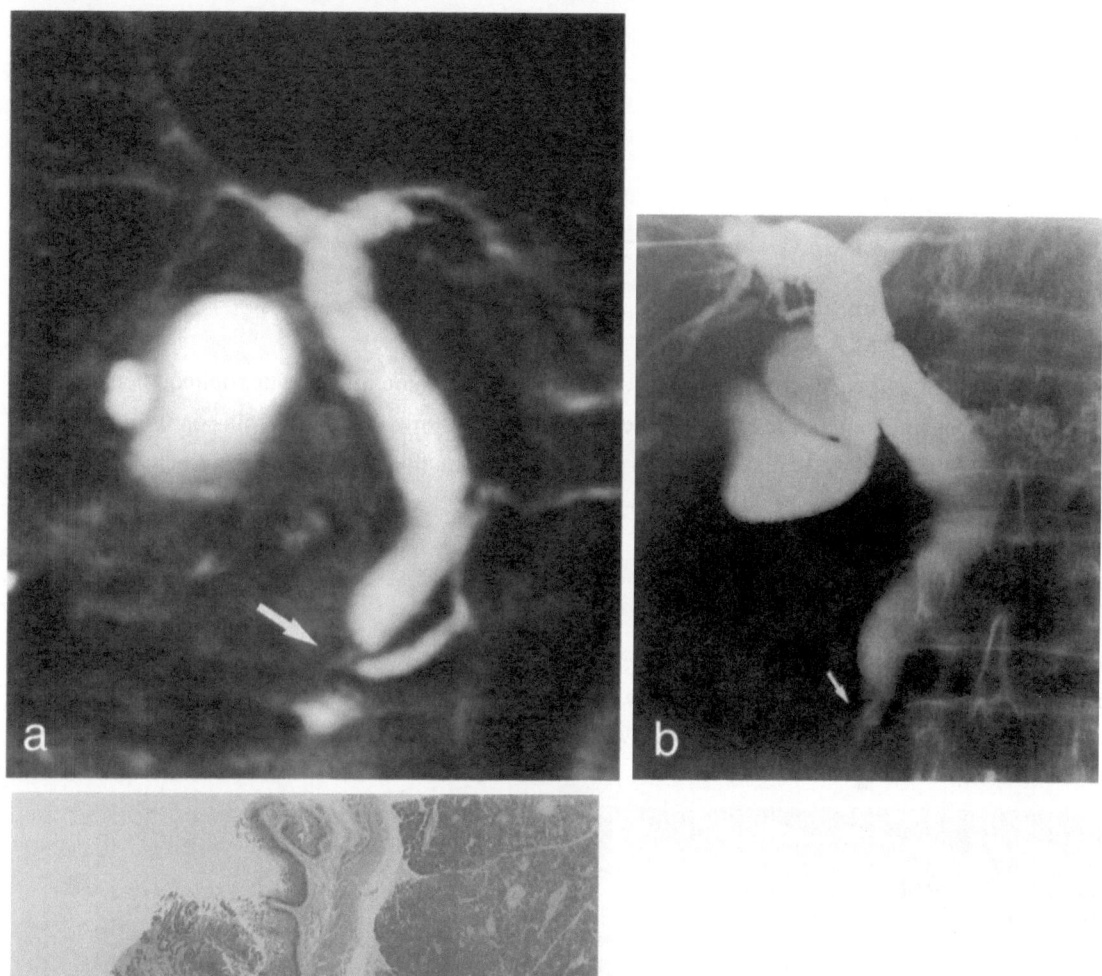

3.9 Benign Stenosis of the Bile Duct

3.9.1 Mirizzi's Syndrome

Mirizzi's syndrome is a state of inflammatory stenosis of the common bile duct or common hepatic duct caused by stones impacting in the neck of the gallbladder or cystic duct [62]. This condition is easily diagnosed by MRCP, which depicts stones in the neck of the gallbladder together with stenosis of the common bile duct or common hepatic duct [63-65].

Fig. 3-34 Mirizzi's syndrome

a. Single-slice MRCP shows a stone impacted in the neck of the gallbladder (*open arrow*), and stenosis of the common hepatic duct (*bold arrow*) is clearly visible. Stones are also observed inside the gallbladder (*arrows*).
b. ERCP shows a stone at the neck of gallbladder (*open arrow*) and stenosis of the common hepatic duct (*arrow*) as in MRCP, but the gallbladder is not opacified.

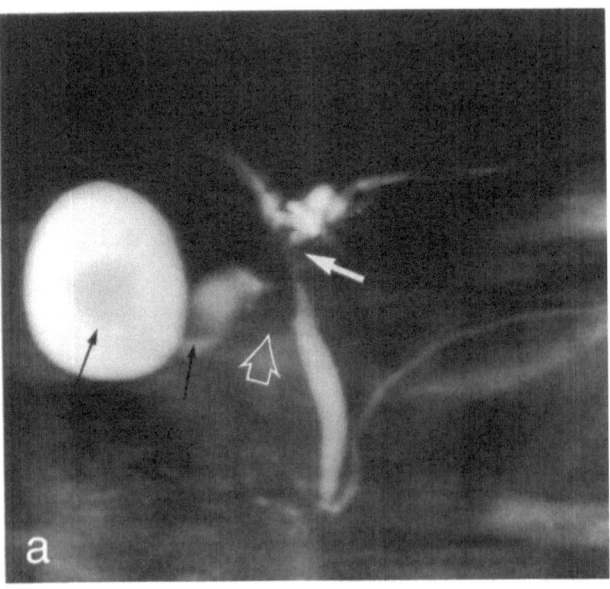
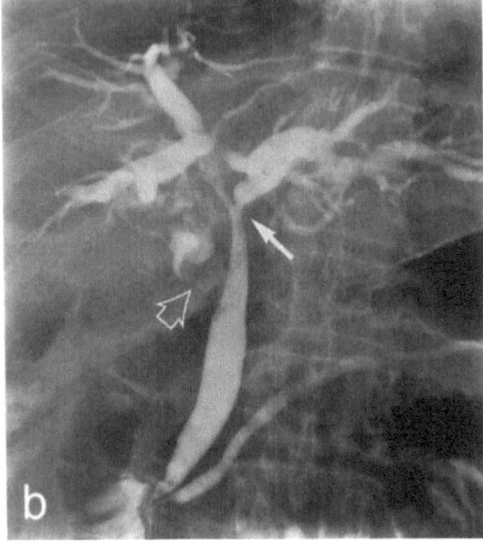

3.9.2   Primary Sclerosing Cholangitis

Primary sclerosing cholangitis (PSC) is a disease of unknown etiology in which multiple segmental strictures occur in both the intrahepatic and extrahepatic bile ducts resulting in chronic cholestasis [66]. Only the intrahepatic ducts are involved in approximately a quarter of the patients [67].

MRCP demonstrates multiple segmental strictures in the intrahepatic and extrahepatic bile ducts, which are characterized as short strictures with "pruned tree" or "beaded" appearance [68]. The isolated duct proximal to an obstruction is clearly delineated on MRCP, but is not opacified with direct cholangiography. Biliary diverticula and webs, which are highly suggestive but not specific findings of PSC, are also visible [69].

Because of its noninvasive nature, MRCP is an effective method to follow up patients with PSC, in which long-term observation is required. By comparison, ERCP usually requires balloon occlusion to adequately depict the intrahepatic bile ducts.

Fig. 3-35 Primary sclerosing cholangitis

a. Three-dimensional MRCP with MIP demonstrates long segments of ductal stenosis in the hilar bile duct (*arrow*). Irregularity of the intrahepatic bile ducts is also depicted.

b. ERCP image obtained after balloon occlusion demonstrates stenosis in the hilar bile duct (*arrow*) and irregularity of the intrahepatic bile ducts. Due to slight ductal distention caused by the pressure of the injected contrast material, the stenoses and irregularity are more clearly delineated on ERCP than on MRCP.

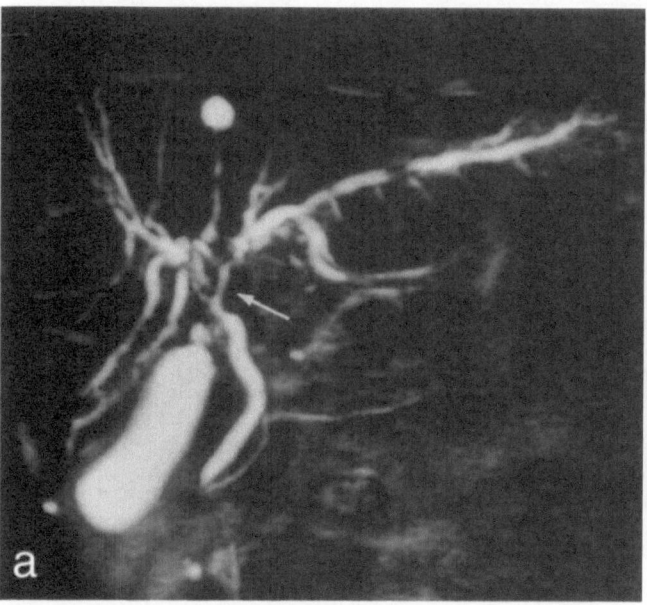
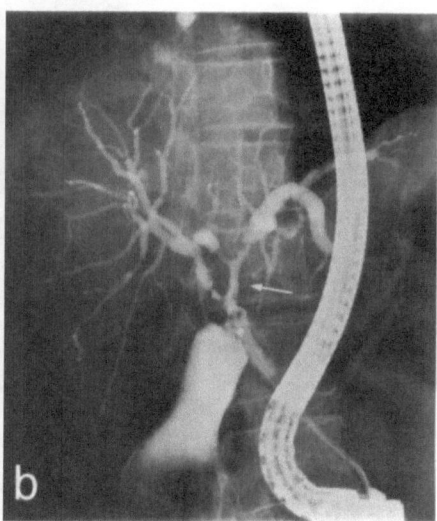

3.9.3   Bile Duct Stenoses Associated with Chronic Pancreatitis

Bile duct stenoses associated with chronic pancreatitis develop as a result of inflammatory fibrosis or edema around the distal common bile duct [70,71]. Pancreatic pseudocysts or abscesses adjacent to the bile duct also may cause the stenoses. These stenoses occur in 9% of patients with chronic pancreatitis [72].

MRCP is useful in the diagnosis of this condition, because stenoses of the bile duct and the pancreatic ductal changes in chronic pancreatitis (see section 4.3.2) are demonstrated simultaneously. Also, the relationship between the common bile duct and pancreatic pseudocyst is clearly depicted. Moreover, there is no risk of exacerbation of the pancreatitis by injecting contrast materials. MRCP demonstrates a long, smooth stenosis of the distal common bile duct, although differentiation from malignant stricture is sometimes difficult [70,71].

Fig. 3-36 Bile duct stenosis associated with chronic pancreatitis

Single-slice MRCP shows a stricture of the intrapancreatic bile duct (*bold arrow*), dilatation of the main pancreatic duct with calculi (*arrowhead*) and cysts in the head of the pancreas (*arrows*). These findings are compatible with bile duct stenosis associated with chronic pancreatitis.

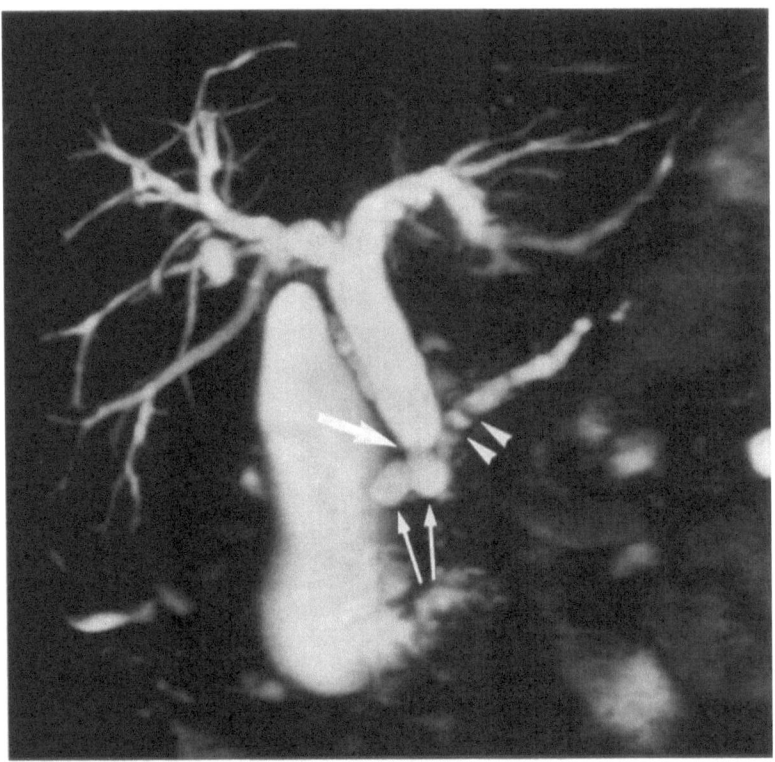

3.9.4   Dysfunction of the Sphincter of Oddi

Dysfunction of the sphincter of Oddi includes both organic and functional alterations, such as stenosis and dyskinesia [73]. Primary fibrosis, glandular hyperplasia, muscle hyperplasia, and muscle hypertrophy are considered to be responsible for the dysfunction [74-77].

MRCP demonstrates a dilated common bile duct that tapers at the level of the ampulla [54]. After secretin injection on MRCP, persistent dilatation of the pancreatic duct is observed in patients with dysfunction of the sphincter of Oddi [78]. However, differentiation from early carcinoma of the papilla of Vater is sometimes difficult [3]. (see section 3.8.6).

Fig. 3-37 Benign stenosis of the sphincter of Oddi

a. Single-slice MRCP shows stenosis of the pancreatobiliary ducts at the level of the ampulla with proximal dilatation (*arrow*). Differentiation from early carcinoma of the papilla of Vater is difficult.
b. ERCP shows the same findings as MRCP (*arrow*).

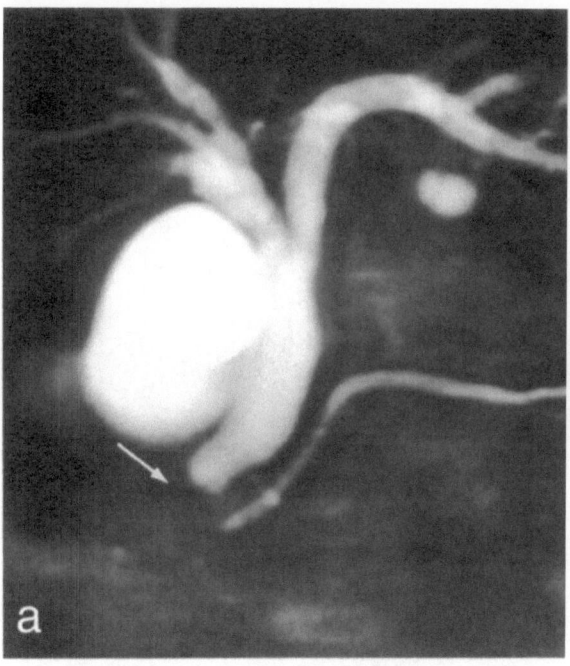
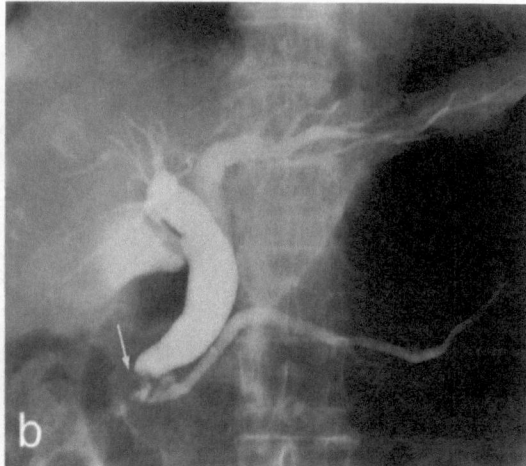

3.9.5   Postoperative Bile Duct Stricture

Iatrogenic focal stenosis of the common bile duct occurs in approximately 0.5% of postcholecystectomy patients [8,79]. This complication is usually recognized months to years after gallbladder surgery when gradual fibrosis and scar formation have caused sufficient luminal narrowing to produce obstructive jaundice [80]. Appropriate interventions are therefore necessary.

MRCP is accurate in depicting the bile duct stricture [3,24,54], and diagnosis can be established easily [81].

Fig. 3-38 Benign fibrous proliferation of the bile duct at the hepatic hilum [82]

a. Single-slice MRCP shows a short stricture in the hilar bile duct (*arrow*) with proximal duct dilatation. Stones are also present in the distal common bile duct (*arrowheads*).
b. Percutaneous transhepatic cholangiography shows a bile duct stricture (*arrow*) and stones (*arrowheads*) as in MRCP.

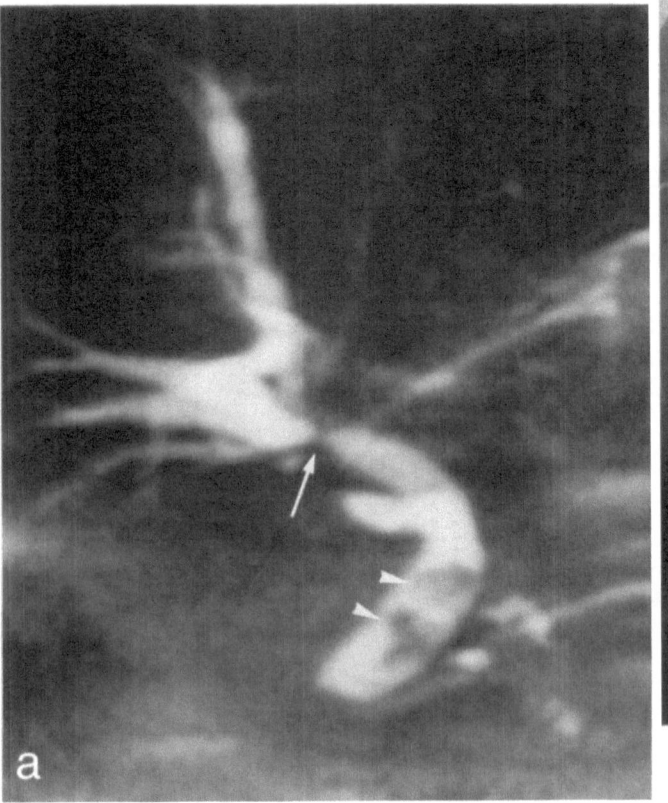
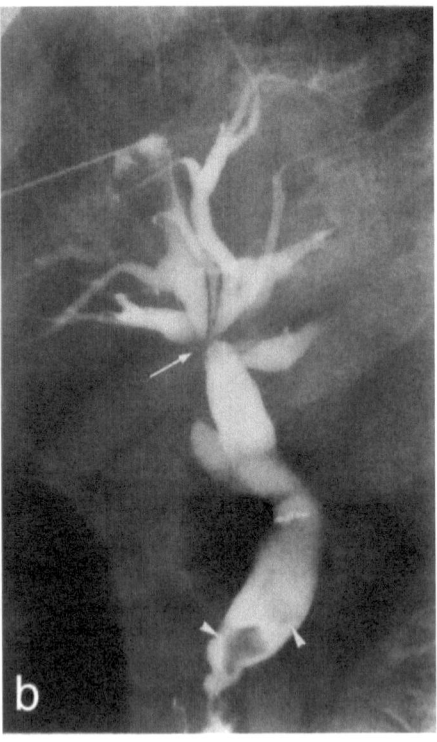

3.10 Biliary-Enteric Anastomoses

Postoperative complications of biliary-enteric anastomoses are relatively common. The incidence of long-term complications, including strictures of the anastomosis, stones, and cholangitis, is reported as 14% [83]. Therefore, postoperative follow-up using MRCP is mandatory.

MRCP depicts the site of the anastomosis, the hepatic ducts, and first- to third-level branches of the intrahepatic bile ducts. Bile duct dilatation, bile duct irregularity caused by cholangitis, anastomotic strictures, and stones are accurately diagnosed by MRCP [38,84].

MRCP is also useful in the diagnosis of the afferent loop syndrome, a known complication of biliary-enteric anastomoses, in which a dilated atonic afferent limb of the jejunum is observed [85].

Furthermore, MRCP after secretin injection is useful to evaluate the anastomotic patency of the pancreatic duct [86].

Fig. 3-39 Postpancreatoduodenectomy evaluation

Single-slice MRCP demonstrates the hepatic duct (*bold arrow*), first- to third-level branches of the intrahepatic bile ducts, pancreatic duct (*arrow*), and the anastomosis site.

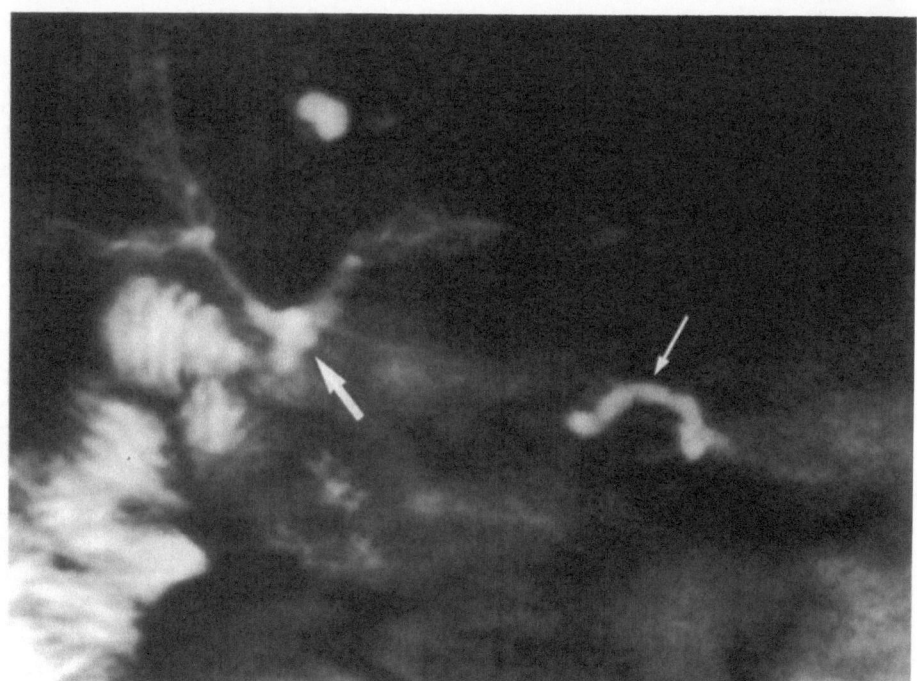

Fig. 3-40 Anastomotic stricture

Single-slice MRCP shows mild stenosis with proximal duct dilatation at the anastomosis of the bile duct (*bold arrow*) and pancreatic duct (*arrow*).

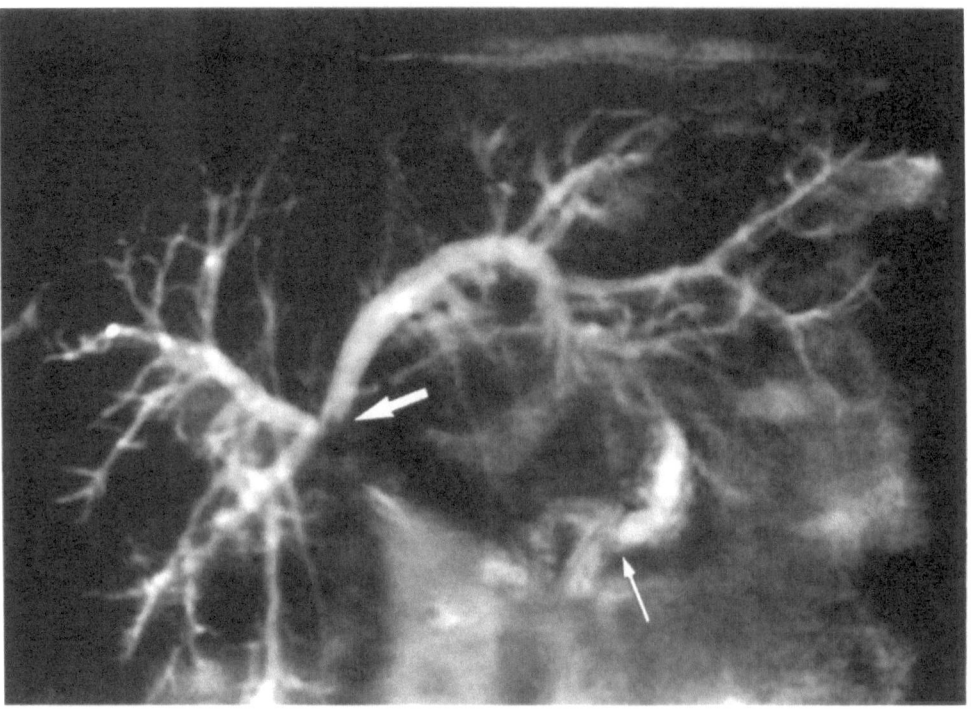

Fig. 3-41 Biliary-enteric anastomosis complicated by stones and cholangitis

Single-slice MRCP demonstrates a stone in the right hepatic duct (*arrow*) and irregularity caused by cholangitis in the intrahepatic bile ducts (*arrowheads*).

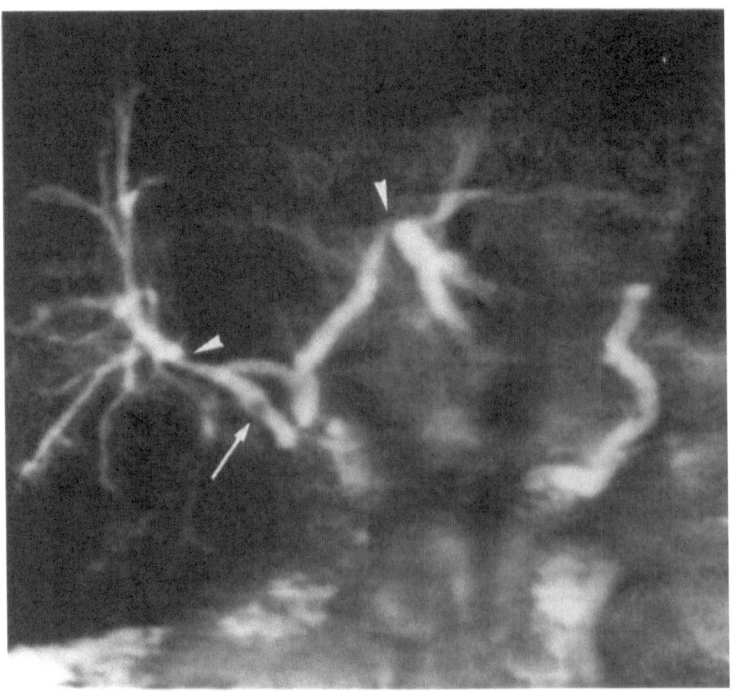

3.11 Other Biliary Tract Conditions

3.11.1 Pneumobilia

Pneumobilia, the presence of air bubbles in the biliary tract, has various causes including biliary-enteric anastomosis, biliary fistula to the gastrointestinal tract, postsphincterectomy, and emphysematous cholangitis. MRCP shows low-intensity areas in the biliary tract, which may mimic stones. Differential diagnosis from calculi is necessary. Axial MR image sections often allow the differentiation of calculi, which lie in the dependent portion of the bile duct lumen, from air bubbles, which float anterior to the bile [25, 87]. When biliary air is present extensively, the biliary tree may not be visualized by MRCP [88] (Fig. 3-15).

Fig. 3-42 Postendoscopic sphincterectomy pneumobilia

Single-slice MRCP shows small low-intensity spots in the common bile duct, indicating pneumobilia.

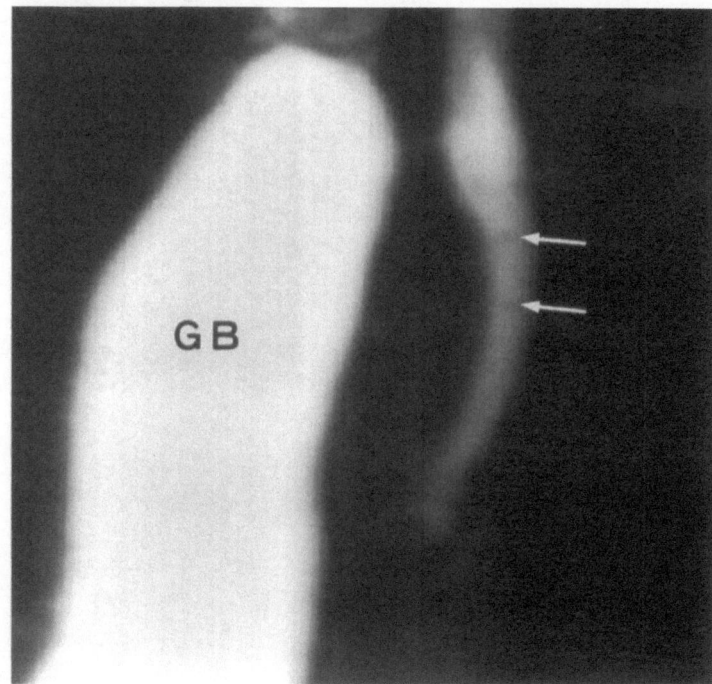

3.11.2   T2-Shortening Bile

On MRCP, the signal intensity of the bile varies in certain conditions including obstructive jaundice (Fig. 3-43), gallstone (Fig. 3-44), cholecystitis, and fasting. A reduction in water content, the presence of extravasated blood, sludge, or proteinaceous components may shorten the T2 value [22,29,89-91].

Fig. 3-43 Obstructive jaundice caused by carcinoma of the papilla of Vater (*arrow*)

Single-slice MRCP shows lower signal intensity of the bile than that of the pancreatic juice (*bold arrow*).

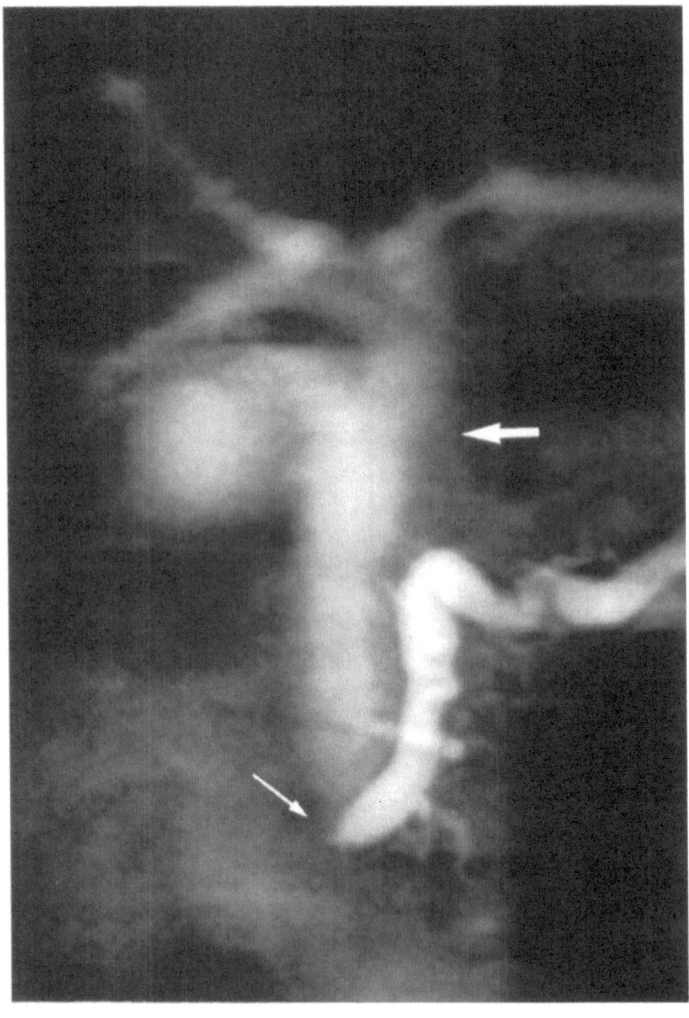

Fig 3-44 Gallstone with cystic duct obstruction and sludge

a. Single-slice MRCP shows an impacted stone in the neck of the gallbladder (*arrow*), and a stone in the fundus (*arrowhead*). The signal intensity inside the gallbladder is lower than that of the bile duct. A small stone is also found in the distal common bile duct.

b. Ultrasonography shows a stone (*arrow*) and sludge in the gallbladder.

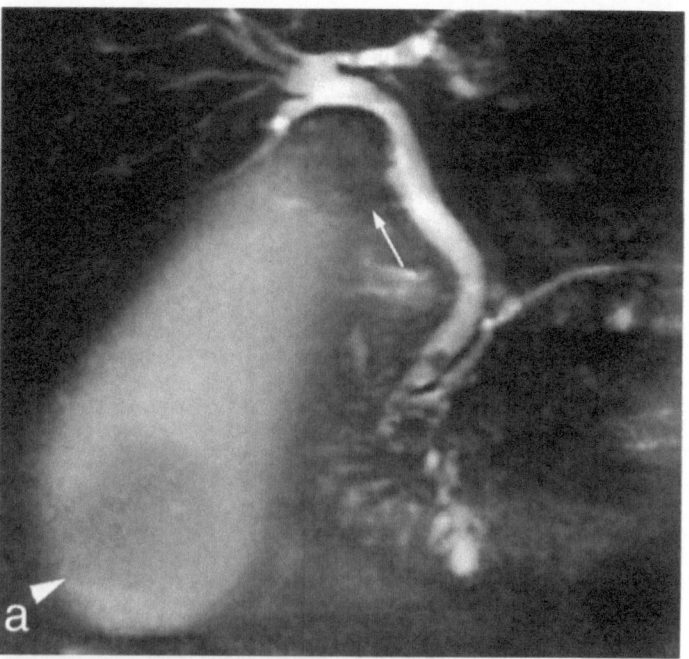

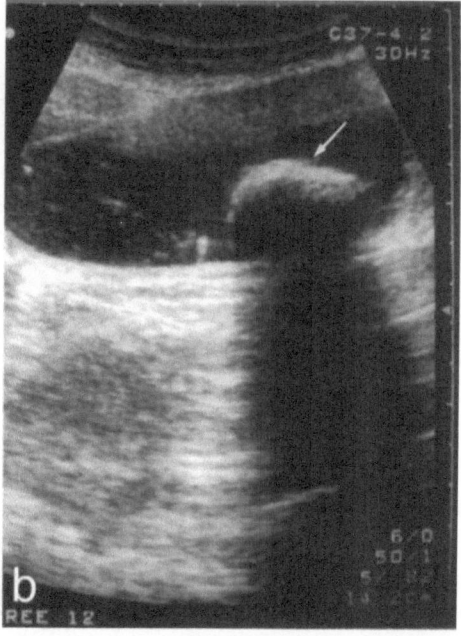

3.11.3   Biliary Stent

MRCP can accurately depict the location and patency of stents, and is useful in follow-up after stent insertion [5,92]. Patency of the stent is indicated by the continuous image of the bile juice inside the tube.

Fig. 3-45 Pancreatic carcinoma after stent insertion

Three-dimensional MRCP with MIP shows a continuous image of bile in the biliary stent (*arrow*). The location and patency of the stent is clearly depicted.

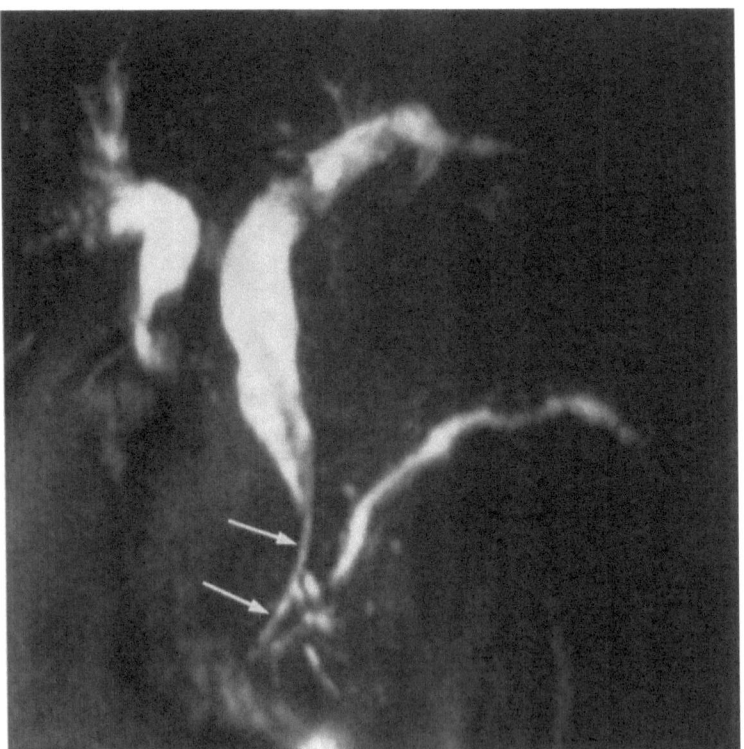

3.11.4   Hepatic Arterial Compression of the Common Hepatic Duct

The right hepatic artery crosses the hepatic duct posterior to the proximal common hepatic duct and may create the appearance of an intraductal filling defect or stenosis [87]. The frequency is reported as 21% [93]. The characteristic location of the hepatic artery has to be noted in the interpretation of images. The right hepatic artery can be confirmed by a flow-sensitive sequence [88].

Fig. 3-46 Hepatic arterial compression of the common hepatic duct

a. Single-slice MRCP shows a pseudostenosis in the proximal common hepatic duct due to compression by the right hepatic artery.
b. ERCP shows no abnormalities in the common hepatic duct.

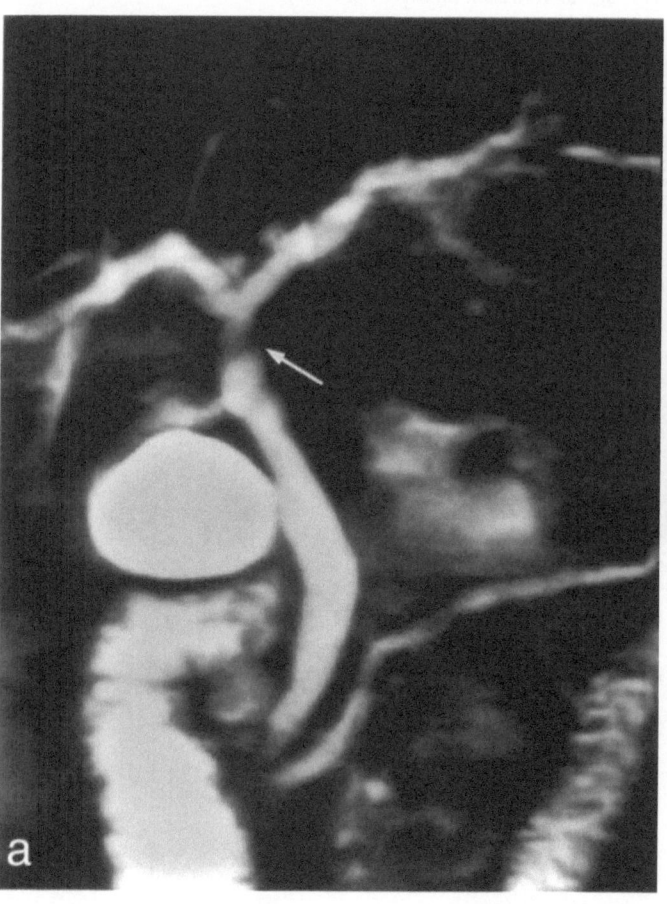
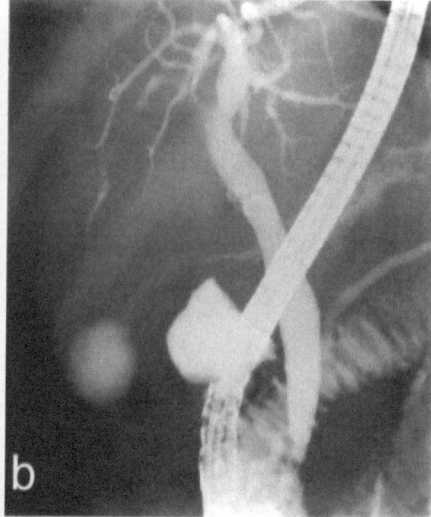

3.11.5   Artifacts Created by Metal Clips

Intra- or extracorporeal devices such as metal clips, metal coils, and other magnetic objects sometimes may produce signal voids. When metal clips are located near the biliary tree, the artifacts may create the appearance of an intraductal filling defect, and differential diagnosis from stones is required. However, with MIP images, careful inspection of the source images usually reveals the eccentric location of the hypointense focus [94]. Plain abdominal radiographs are helpful in confirming the clips.

Fig. 3-47 Artifacts created by a surgical clip

a. Single-slice MRCP shows a signal void in the proximal common hepatic duct (*arrow*).

b. ERCP shows a surgical clip at the same site as the signal void seen on MRCP (*arrow*).

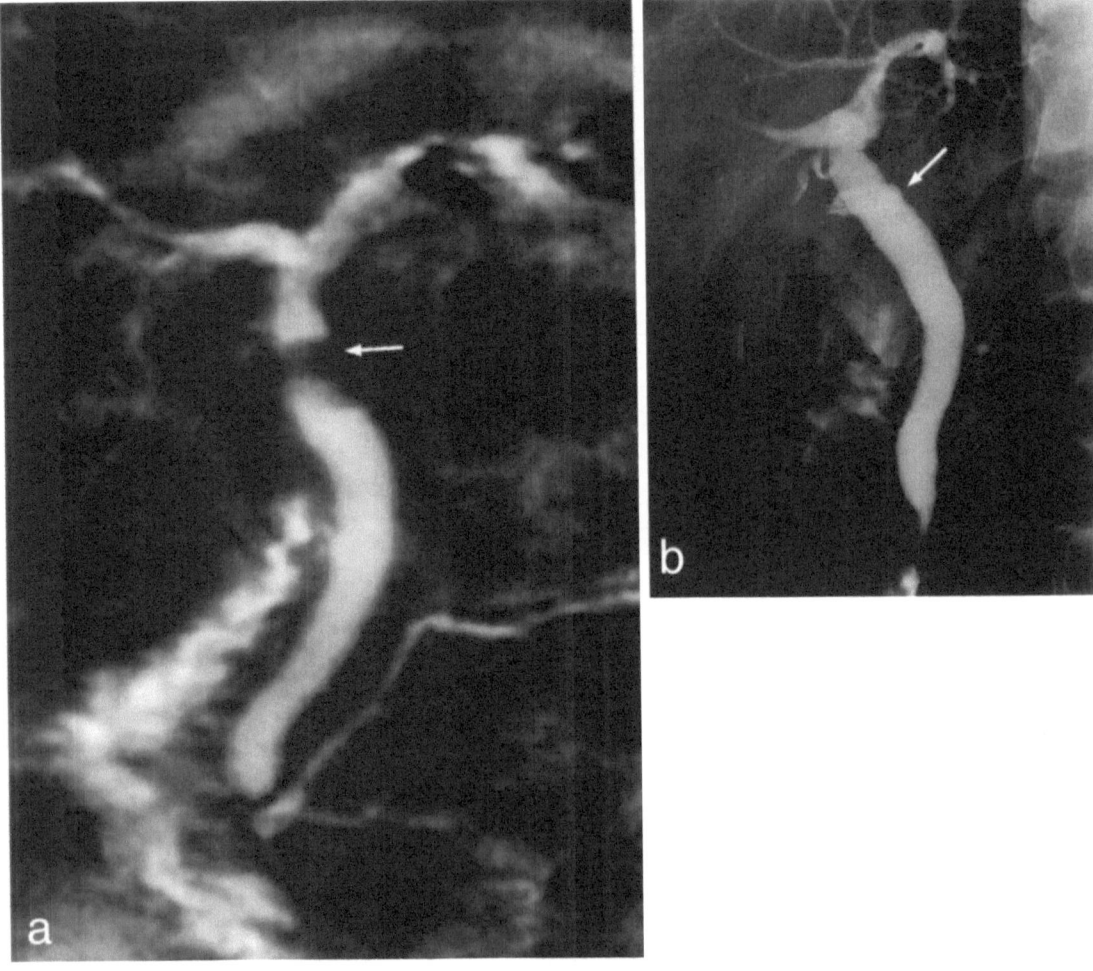

References

1. Miyazaki K, Yamashita Y, Tsuchigame T, et al (1996) MR cholangiopancreatography using HASTE (half-Fourier acquisition single-shot turbo spin-echo) sequences. AJR Am J Roentgenol 166: 1297-1303

2. Laubenberger J., Buchert M, Shneider B, et al (1995) Breath-hold projection magnetic resonance cholangiopancreatography (MRCP): a new method for the examination of the bile duct and pancreatic ducts. Magn Reson Med 33:18-23

3. Sai J (1997) MR Cholangiopancreatography in the diagnosis of biliary tract carcinoma at an early stage (in Japanese). J Jpn Bil Assoc 11:151-157.

4. Taourel P, Bret PM, Reinhold C, et al (1996) Anatomic variants of the biliary tree: diagnosis with MR cholangiopancreatography. Radiology 199:521-527

5. Sai J, Asahara S, Ariyama J, et al (1997) New diagnostic procedure of the biliary disease: MRCP, IDUS (in Japanese). Geka 59:259-266

6. Puente SG, Bannura GC (1983) Radiological anatomy of the biliary tract: variations and congenital abnormalities. World J Surg 7:271-276

7. Shaw MJ, Dorsher PJ, Vennes JA (1993) Cystic duct anatomy: an endoscopic perspective. Am J Gastroenterol 88:2102-2106

8. Deziel DJ, Millikan KW, Economou SG, et al (1993) Complications of laparoscopic cholecystectomy: a national survey of 4,292 hospitals and an analysis of 77,604 cases. Am J Surg 165:9-14

9. Sai J, Ariyama J, Suyama M, et al (1998) 3D-MRCP (in Japanese). J Bil Pancr 19:197-203

10. Newcombe JF, Henley FA (1964) Left-sided gallbladder, A review of the literature and a report of a case associated with hepatic duct carcinoma. Arch Surg 88:494-497

11. Sai J, Ariyama J (1997) MR cholangiopancreatography in the diagnosis of pancreaticobiliary maljunction. In: Komi N, Funabiki T (eds) Pancreaticobiliary maljunction:s consensus and controversy. Igakutosho, Tokyo, pp 58-62

12. Van Hoe L, Gryspeerdt S, Vanbeckevoort D, et al (1998) Normal Vaterian sphincter complex: evaluation of morphology and contractibility with dynamic single-shot MR cholangiography. AJR Am J Roentgenol 170:1497-1500

13. Babitt DP (1969) Congenital choledochal cyst: new etiology concept based on anomalous relationships of the common bile duct and pancreatic bulb. Ann Radiol 12: 231-240

14. Kimura K, Ohto M, Saisho H, et al (1985) Association of gallbladder carcinoma and anomalous pancreatobiliary ductal union. Gastroenterology 89:1258-1265

15. Committee of JSPBM for Diagnostic Criteria (1994) Diagnostic criteria of pancreatico-biliary maljunction. J Hep Bil Panc Surg 1:219-221

16. Komi N, Tamura T, Miyoshi Y, et al (1984) Nationwide survey of cases of choledochal cyst. Analysis of coexistent anomalies, complications and surgical treatment in 645 cases. Surg Gastroenterol 3:69-73

17. Aoki H, Sugatani H, Shimizu M (1987) A clinical study on cancer of the bile duct associated with anomalous arrangements of pancreaticobiliary ductal system. Analysis of 569 cases collected in Japan (in Japanese). J Bil Tract Pancr 8:1539-1551

18. Hirohashi S, Hirohashi R, Uchida H, et al (1997) Pancreatitis: evaluation with MR cholangiopancreatography in children. Radiology 203:411-415

19. Irie H, Honda H, Jimi M, et al (1998) Value of MR cholangiopancreatography in evaluating choledochal cyst. AJR Am J Roentgenol 171:1381-1385

20. Matos C, Nicaise N, Deviere J, et al (1998) Choledocal cysts: comparison of findings at MR cholangiopancreatography and endoscopic retrograde cholangiopancreatography in eight patients. Radiology 209:443-448

21. Sugiyama M, Baba M, Atomi Y, et al (1998) Diagnosis of anomalous pancreaticobiliary junction: Value of magnetic resonance cholangiopancreatography. Surgery 123:391-397

22. Reuther G, Kiefer B, Tuchmann A (1996) Cholangiography before biliary surgery: single shot MR Cholangiopancreatography versus intravenous cholangiography. Radiology 198:561-566

23. Reinhold C, Taourel P, Bret PM, et al (1998) Choledocholithiasis: Evaluation of MR cholangiography for diagnosis. Radiology 209:435-442

24. Fulcher AS, Turner MA, Capps GW, et al (1998) Half-Fourier RARE MR cholangiopancreatography: experience in 300 subjects. Radiology 207:21-32

25. Bret PM, Reinhold C (1997) Magnetic resonance cholangiopancreatography Endoscopy 29:472-486

26. Moon KL Jr, Hricak H, Margulis AR, et al (1983) Nuclear magnetic resonance imaging characteristics of gallstones in vitro. Radiology 148:753-756

27. Baron EL, Shuman WP, Lee SP, et al (1989) MR appearance of gallstones in vitro at 1.5T: correlation with chemical composition. AJR Am J Roentgenol 153:497-502

28. Chan YI, Chan AC, Lam WW, et al (1996) Cholelithiasis: comparison of MR cholangiography and endoscopic retrograde cholangiography. Radiology 200:85-89

29. Park MS, Yu JS, Kim YH, et al (1998) Acute cholecystitis: comparison of MR cholangiography and US. Radiology 209:781-785

30. Regan F, Fradin J, Khazan R, et al (1996) Choledocholithiasis: evaluation with MR cholangiopancreatography. AJR Am J Roentgenol 167:1441-1445

31. Schoenfield LJ, Cary MC, Marks JW, et al (1989) Gallstones: an update. Am J Gastroenterol 84:999-1007

32. Tanimura H. Uchiyama K, Ishiyama K, et al (1994) Epidemiology of Hepatolithiasis (in Japanese). JBil Tract Pancr 15:401-408

33. Chen MF, Jan YY, Wang CS, et al (1989) Intrahepatic stones associated with cholangiocarcinoma. Am J Gastroenterol 84:391-395

34. Kawarada Y, Mita T (1994) Cholangiocarcinoma associated with hepatolithiasis (in Japanese). J Bil Tract Pancr 15:435-446

35. Kubo S, Hamba H, Hirohashi K, et al (1997) Magnetic resonance cholangiography in hepatolithiasis. Am J Gastroenterol 92:629-632

36. Berk JE (1940) Management of acute cholecystitis. Am J Dig Dis 7:325-332

37. Glenn F (1976) Acute cholecystitis. Surg Gynecol Obstet 143:56-60

38. Sai J, Ariyama J, Suyama M, et al (1998) Clinical usage of MR cholangiopancreatography (in Japanese). Nippon Rinsho 56:2768-2772

39. Regan F, Schaefer DC, Smith DP, et al (1998) The diagnostic utility of HASTE MRI in the evaluation of acute cholecystitis. Half-Fourier acquisition single-shot turbo SE. J Comput Assist Tomogr 22:638-642

40. Ralls PW, Colleti PM, Lapin SA, et al (1985) Real-time sonography in suspected acute cholecystitis. Radiology 155:767-771

41. Laing FC, Federle MP, Jefferey RB, et al (1981) Ultrasonic evaluation of patients with acute right upper quadrant pain. Radiology 140:449-455

42. Jutras JA, Aongtin JM, Levesque MD (1960) Hyperplastic cholecystoses. Hickey Lecture. Am J Roentgenol 83:795-827

43. Fotopoulos JP, Crampton AR (1964) Adenomyomatosis of the gallbladder. Med Clin N Am 48:9-36

44. Williams I, Slavin G, Cox A, et al (1986) Diverticular disease (adenomyomatosis) of the gallbladder: a radiological-pathological survey. Br J Radiol 59:29-34

45. Beilby JO (1967) Diverticulosis of the gall bladder. The fundal adenoma. Br J Exp Pathol 48:455-461

46. Ootani T, Shirai Y, Tsukada K, et al (1992) Relationship between gallbladder carcinoma and the segmental type of adenomyomatosis of the gallbladder. Cancer 69:2647-2652

47. Sai J, Ariyama J, Suyama M (1997) MR cholangiopancreatography in the diagnosis of anomalous union of the pancreaticobiliary ducts (in Japanese). Kan Tan Sui 35: 231-235

48. Yoshimitsu K, Honda H, Jimi M, et al (1999) MR diagnosis of adenomyomatosis of the gallbladder and differentiation from gallbladder carcinoma: importance of showing Rokitansky-Aschoff sinuses. AJR Am J Roentgenol 172:1535-1540

49. Halpert RD, Bedi DG, Tirman PJ, et al (1989) Segmental adenomyomatosis of the gallbladder. Am Surg 55:570-572

50. van der Vegt JH, Berk RN, Lichtenstein JE (1983) The hyperplastic cholestoses: cholesterolosis and adenomyomatosis. Radiology 146:593-601

51. Koga A, Watanabe K, Fukuyama T, et al (1998) Diagnosis and operative indications for polypoid lesions of the gallbladder. Arch Surg 123:26-29

52. Nagaiwa J (1992) Clinico-pathological study of polypoid lesion of the gallbladder (in Japanese). J Jpn Bil Ass 6:371-379

53. Chijiiwa K, Sumiyoshi K, Nakayama F (1991) Impact of recent advances in hepatobiliary imaging techniques on the preoperative diagnosis of carcinoma of the gallbladder. World J Surg 15:322-327

54. Guibaud L, Bret PM, Reinhold C, et al (1995) Bile duct obstruction and choledocholithiasis: diagnosis with MR cholangiopancreatography. Radiology 197:109-115

55. Sai J, Ariyama J, Suyama M, et al (1996) Breath-hold MR cholangiopancreatography with fast advanced spin echo technique in the diagnosis of malignant obstruction of the lower biliary tract. HPB Surg 9:433

56. Mizumoto R, Ogura Y, Matsuda N, et al (1990) Cooperative survey of surgical treatment for carcinoma of the biliary tract in Japan (in Japanese). J Bil Tract Pancr 11:869-882

57. Sai J, Ariyama J, Suyama M, et al (1996) Early diagnosis of extrahepatic bile duct cancer with MR cholangiopancreatography. Gut 39:A184

58. Nimura Y, Hayakawa N, Kamiya J, et al (1995) Hilar cholangiocarcinoma — surgical anatomy and curative resection. J Hep Bil Pancr Surg 2:239-248

59. Fulcher AS, Turner MA (1997) HASTE MR cholangiopancreatography in the evaluation of hilar cholangiocarcinoma. AJR Am J Roentgenol 169:1501-1505

60. Semelka RC, Kelekis NL, John G, et al (1997) Ampullary carcinoma: demonstration by current MR techniques. J Magn Reson Imaging 7:153-156

61. Darweesh RMA, Thorsen MK, Dodds WJ, et al (1988) Computed tomography examination of periampullary neoplasms. J Comput Assist Tomogr 12:36-41

62. Mirizzi PL (1948) Sindrome del conducto hepatico. J Int Chir 8:731-777

63. Clemmett AR, Lowman RM (1965) The roentgen feature of the Mirizzi syndrome. AJR Am J Roentgenol 94:480-483

64. Sai J, Ariyama J (1999) MRCP in the diagnosis of pancreatobiliary diseases: its progression and limitation. Jpn J Gastroenterol 96:259-265

65. Becker CD, Grossholz M, Becker M, et al (1997) Choledocholithiasis and bile duct stenosis: diagnostic accuracy of MR cholangiopancreatography. Radiology 205:523-530

66. La Russo NF, Wiesner RH, Ludwig J, et al (1984) Primary sclerosing cholangitis. N Engl J Med 310:899-903

67. Stockbr‚gger RW, Olsson R, Jaup B, et al (1988) Forty-six patients with primary sclerosing cholangitis. Radiological bile duct changes in relationship to clinical course and concomitant inflammatory bowel disease. Hepatogastroenterology 35:289-294

68. MacCarty RL, LaRusso NF, Wiesner RH, et al (1983) Primary sclerosing cholangitis: findings on cholangiography and pancreatography. Radiology 149:39-44

69. Gulliver DJ, Baker ME, Putnam W, et al (1991) Bile duct diverticula and webs: nonspecific cholangiographic features of primary sclerosing cholangitis. AJR Am J Roentgenol 157:281-285

70. Rohrmann CA Jr, Baron RL (1989) Biliary complications of pancreatitis. Radiol Clin N Am 27: 93-104

71. Sarles H (1978) Cholestasis and lesions of the biliary tract in chronic pancreatitis. Gut 19: 851-857

72. Petrozza JA, Dutta SK, Latham PS, et al (1984) Prevalence and natural history of distal common bile duct stenosis in alcoholic pancreatitis. Dig Dis Sci 29: 890-895

73. Corazziari E, Funch-Jensen P, Hogan WJ, et al (1993) Functional disorders of the biliary tract. Gastroenterol Int 6:129-144

74. Colcock BP (1958) Stenosis of the sphincter of Oddi. Surg Clin N Am 38:631

75. Catell RB, Colcock BP, Pollack JL (1957) Stenosis of the sphincter of Oddi. N Engl J Med 256:429

76. Manier JW, Cohen WN, Printen KJ (1974) Dysfunction of the sphincter of Oddi in a postcholecystectomy patient. Am J Gastroenterol 62:148-150

77. Paulino F, Cavalcanti A (1960) Anatomy and pathology of the distal common bile duct. Special reference to stenotic odditis. Am J Dig Dis 5:697-713

78. Matos C, Metens T, Deviere J, et al (1997) Pancreatic duct: morphologic and functional evaluation with dynamic MR pancreatography after secretin stimulation. Radiology 203:435-441

79. Ghahremani GG, Crampton AR, Benstein JR, et al (1991) Iatrogenic biliary tract complications: radiologic features and clinical significance. Radiographics 11:441-456

80. Ghahremani GG (1994) Postsurgical and traumatic lesions of the biliary tract. In: Gore RM, Levine MS, Laufer I (eds) Text book of gastrointestinal radiology. Saunders, Philadelphia, pp 1762-1778

81. Coakley FV, Schwartz LH, Blumgart LH, et al (1998) Complex postcholecystectomy biliary disorders: preliminary experience with evaluation by means of breath-hold MR cholangiography. Radiology 209:141-146

82. Hadjis NS, Collier NA, Blumgart LH (1985) Malignant masquerade at the hilum of the liver. Br J Surg 72:659-661

83. Bismuth H, Franco D, Corlette MB, et al (1978) Long term results of Roux-en-Y hepaticojejunostomy. Surg Gynecol Obstet 146:161-167

84. Pavone P, Laghi A, Catalano C, et al (1997) MR cholangiography in the examination of patients with biliary-enteric anastomoses. AJR Am J Roentgenol 169:807-811

85. Mckee JD, Raju GPR, Edelman RR, et al (1997) MR cholangiopancreatography (MRCP) in diagnosis of afferent loop syndrome presenting as cholangitis. Dig Dis Sci 42:2082-2086

86. Sho M, Nakajima Y, Kanehiro H, et al (1998) A new evaluation of pancreatic function after pancreatoduodenectomy using secretin magnetic resonance cholangiopancreatography. Am J Surg 176:279-282

87. Fulcher AS, Turner MA (1998) Pitfalls of MR cholangiopancreatography. J Comput Assist Tomogr 22:845-850

88. David V, Reinhold C, Hochman M, et al (1998) Pitfalls in the interpretation of MR cholangiopancreatography. AJR Am J Roentgenol 170:1055-1059

89. Hricak H, Filly RA, Margulis AR, et al (1983) Work in progress: nuclear magnetic resonance imaging of the gallbladder. Radiology 147:481-484

90. Demas BE, Hricak H, Moseley M, et al (1985) Gallbladder bile : an experimental study in dogs using MR imaging and proton MR spectroscopy. Radiology 157: 453-455

91. Loflin TG, Simeone JF, Mueller PR, et al (1985) Gallbladder bile in cholecystitis: in vitro MR evaluation. Radiology 157:457-459

92. Soto JA, Barish MA, Yucel EK, et al (1996) Magnetic resonance cholangiopancreatography: comparison with endoscopic retrograde cholangiopancreatography. Gastroenterology 110:589-597

93. Kondo H, Kanematsu M, Shiratori Y, et al (1999) Potential pitfall of MR cholangiopancreatography: right hepatic artery impression of the common hepatic duct. J Comput Assist Tomogr 23:60-62

94. Soto JA, Barish MA, Yucel EK, et al (1995) MR cholangiopancreatography: findings on 3D fast spin-echo imaging. AJR Am J Roentgenol 165:1397-1401

Chapter 4 The Pancreas

4.1 Normal Pancreatic Duct

MRCP using the single-shot fast spin echo techniques (RARE, HASTE, FASE, SSFSE) allows routine visualization of the normal pancreatic duct system, including the main pancreatic duct, Santorini's duct, and side branches in the pancreatic head [1].

In our study of 30 normal subjects, MRCP depicted the normal pancreatic duct system at a rate of 100% for the main pancreatic duct, 93% for Santorini's duct, and 83% for side branches in the pancreatic head [2]. However, the normal nondilated side branches in the body and tail are usually not visible.

MRCP depicts the physiological state of the pancreatic duct. The duct caliber on ERCP may be larger than that seen on MRCP, due to the pressure of the injected contrast material in ERCP [3].

Fig. 4-1 Normal pancreatic duct system

a. Single-slice MRCP shows the main pancreatic duct, Santorini's duct (*arrowhead*), and the side branches in the pancreatic head (*arrow*).

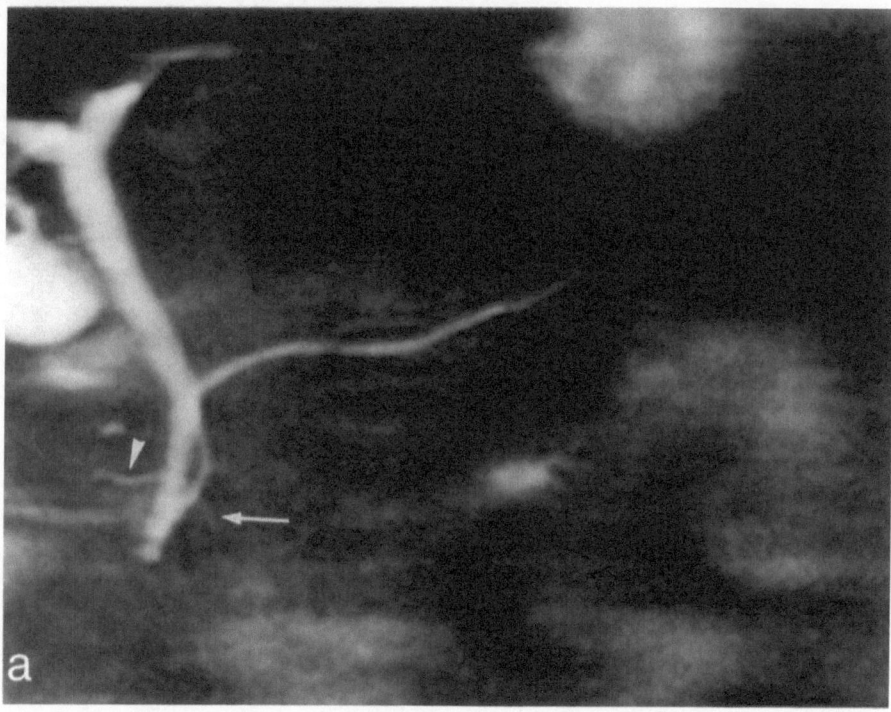

Fig. 4-1   *continued*

b. ERCP of the same case shows the
pancreatic duct system similar to
depiction by MRCP.

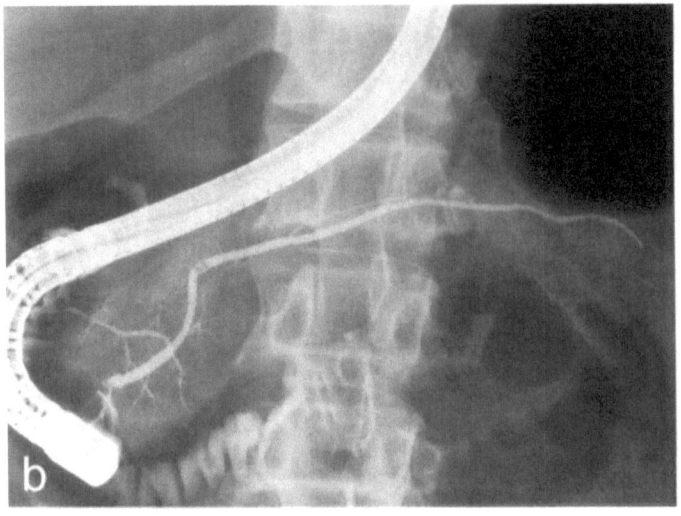

Fig. 4-2 Normal pancreatic duct system

Single-slice MRCP of the right anterior oblique plane (Fig. 2-1) clearly shows the pancreatic duct in the tail (*arrow*).

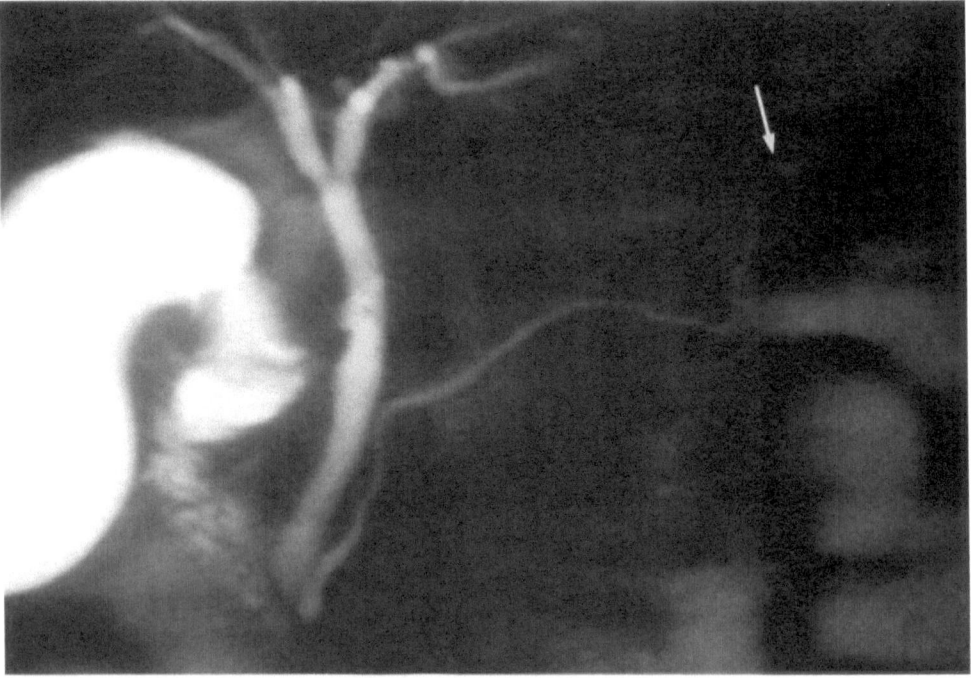

4.2 Pancreas Divisum

Pancreas divisum is a congenital anomaly in which the ventral and the dorsal pancreatic anlagen fail to fuse. The prevalence of this anomaly is reported to be 4%–10% [4,5]. In cases of complete divisum, the absence of fusion is complete, while in cases of incomplete divisum, there is a small connection between the ventral and dorsal pancreatic ducts.

MRCP permits simultaneous visualization of the dorsal and ventral pancreatic ducts [6,7], and its accuracy is reported to be same as that for ERCP [6]. The characteristic sign on MRCP is that the larger dominant dorsal pancreatic duct can be seen in the head of the pancreas, crossing anteriorly to the common bile duct and draining superiorly and separately from the common bile duct [7]. However, it is often difficult to differentiate between complete and incomplete pancreas divisum [6].

On ERCP, the dorsal pancreatic duct is often not visualized due to difficulty of cannulation into the accessory papilla. When imaging is limited to the ventral pancreatic duct, differentiation is required from pancreatic carcinomas and congenital aplasia of the body and tail of the pancreas.

Fig. 4-3 Pancreas divisum

Single-slice MRCP shows the short, thin, blind-ended ventral duct (*arrow*), which does not communicate with the dorsal duct (*arrowheads*).

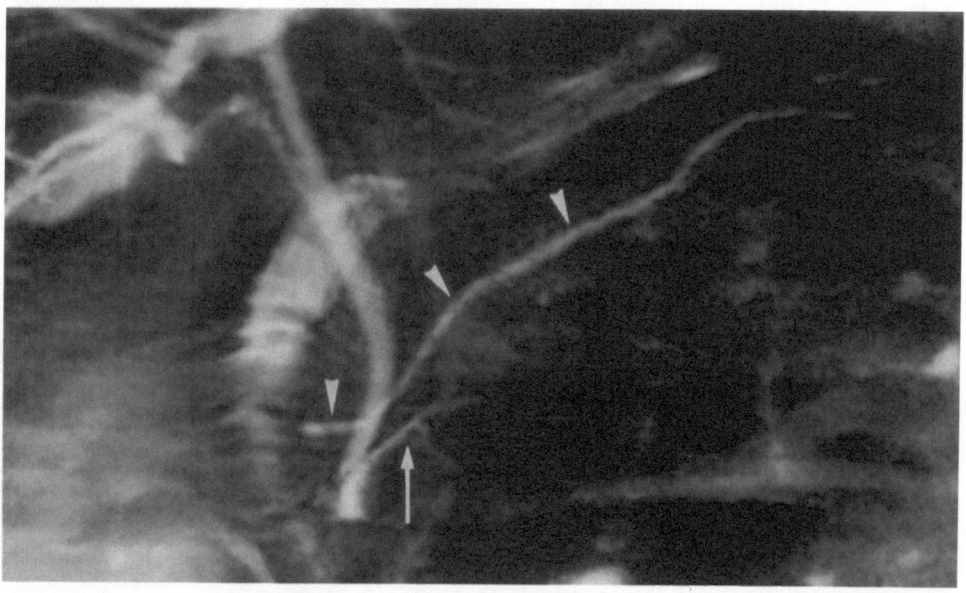

4.3 Pancreatitis

4.3.1 Acute Pancreatitis

Because MRCP detects water signals with high sensitivity, it can depict inflammatory fluid collection associated with acute pancreatitis more clearly than can US and CT [8]. Therefore, MRCP not only depicts the pancreatic duct and biliary tract, but also peripancreatic fluid collection, or pancreatic pseudocysts associated with acute pancreatitis [9]. Underlying pathologic conditions, such as biliary calculi, stenosis of the pancreatic duct, stones in the pancreatic duct, the presence of chronic pancreatitis, pancreas divisum [10], and pancreaticobiliary maljunction [11], are easily diagnosed. MRCP is the preferred modality, for its

Fig. 4-4 Acute pancreatitis

Single-slice MRCP shows pseudocysts (*arrows*) and peripancreatic effusion (*arrowheads*) in the body and tail of the pancreas. Pleural effusion (*bold arrows*) is also depicted.

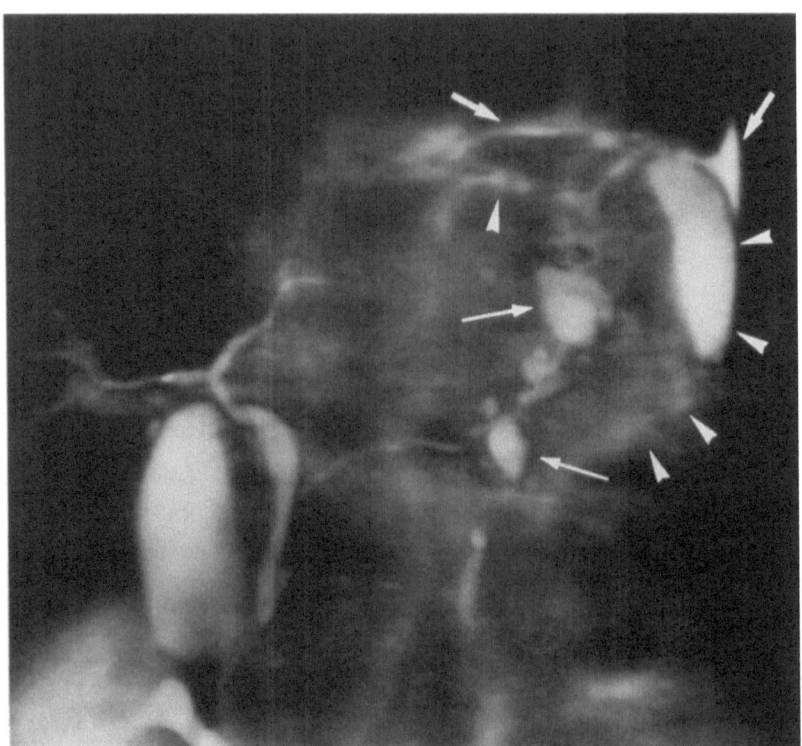

noninvasive nature, in diagnosing acute pancreatitis, and ERCP is indicated only for those patients in whom therapeutic intervention is required. Also, MRCP is useful to observe the clinical course of acute pancreatitis and the progress of size and number of pseudocysts.

4.3.2   Chronic Pancreatitis

MRCP is useful in depicting dilated or narrowed segments of the main pancreatic duct, which are hallmarks of chronic pancreatitis [3,12,13]. Dilated segments of the pancreatic duct on MRCP are underestimated, and narrowed segments are overestimated compared with those seen on ERCP, due to the pressure of the injected contrast material in ERCP [3]. Pancreatic calculi or protein plugs appear as intraductal filling defects in the main pancreatic duct. In previous reports, the sensitivity of MRCP using fast-spin echo technique was 83%–100% for dilatation, 70%–92% for narrowing, and 92%–100% for filling defect [3,14].

MRCP also can depict the subtle side-branch dilatation and small cavities that indicate the early changes of chronic pancreatitis [15]. In advanced chronic pancreatitis, the main pancreatic duct is depicted with a "chain of lakes" appearance.

In addition, MRCP can detect complications of chronic pancreatitis, including pseudocysts, and bile duct stenosis (see section 3.9.3).

Table 4-1 Classification of pancreatograms in chronic pancreatitis

Terminology	Main duct	Abnormal side branches	Additional features
Normal	Normal	None	
Equivocal	Normal	Fewer than 3	
Mild changes of chronic pancreatitis	Normal	3 or more	
Moderate changes of chronic pancreatitis	Abnormal	More than 3	
Marked changes of chronic pancreatitis	Abnormal	More than 3	One or more of: large cavity, obstruction, filling defects, severe dilatation or irregularity

Fig. 4-5 Chronic pancreatitis

Single-slice MRCP shows irregular and tortuous dilatation of the main pancreatic duct. Dilatation of side branches is also delineated (*arrowheads*). Small filling defects corresponding to calculi (*arrows*) are seen in the main pancreatic duct.

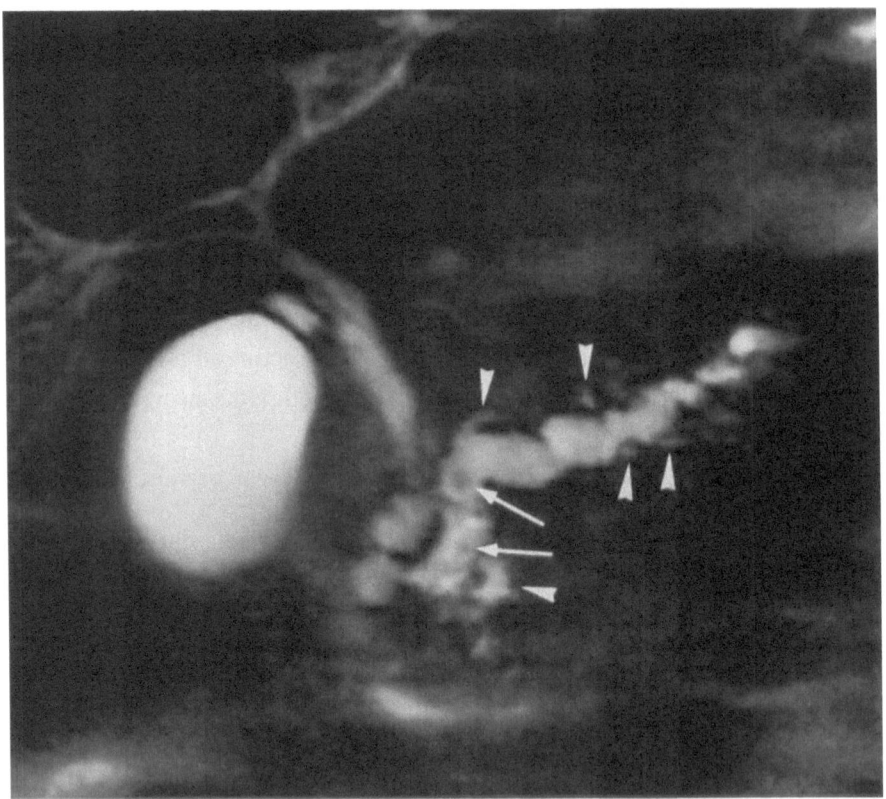

Fig. 4-6 Chronic pancreatitis

a. Single-slice MRCP shows mild dilatation of the main pancreatic duct and slightly dilated side branches (*arrowheads*). Retention cysts (*arrows*) associated with chronic pancreatitis are also depicted.

b. ERCP shows the same findings as MRCP, except that side branches are more clearly depicted and the retention cysts are not visible.

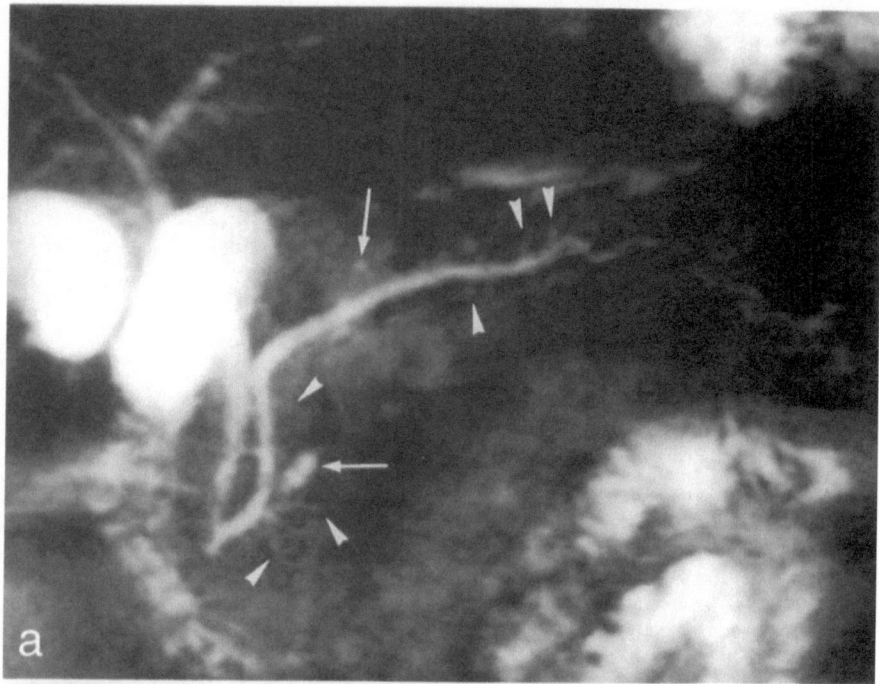

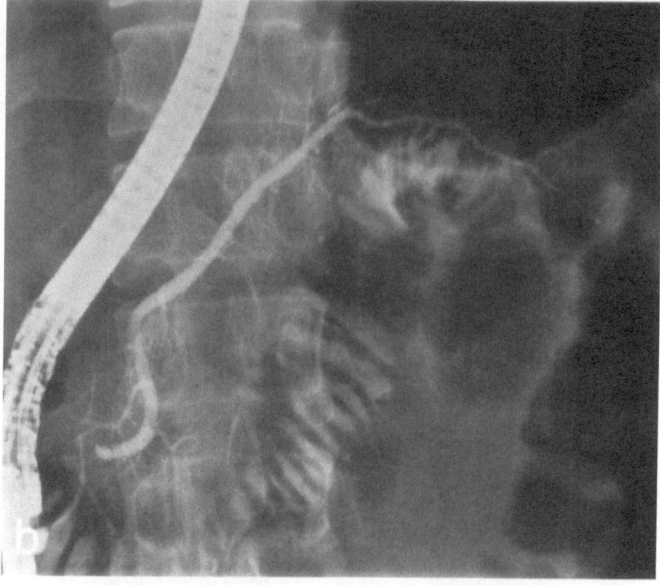

Fig. 4-7 Chronic pancreatitis

a. Single-slice MRCP shows dilated or narrowed segments of the main pancreatic duct and mild dilatation of side branches (*arrowheads*). Retention cysts (*arrows*) associated with chronic pancreatitis also are depicted.

b. ERCP shows the same findings as MRCP, except that the side branches are more clearly depicted and the retention cysts are not visible.

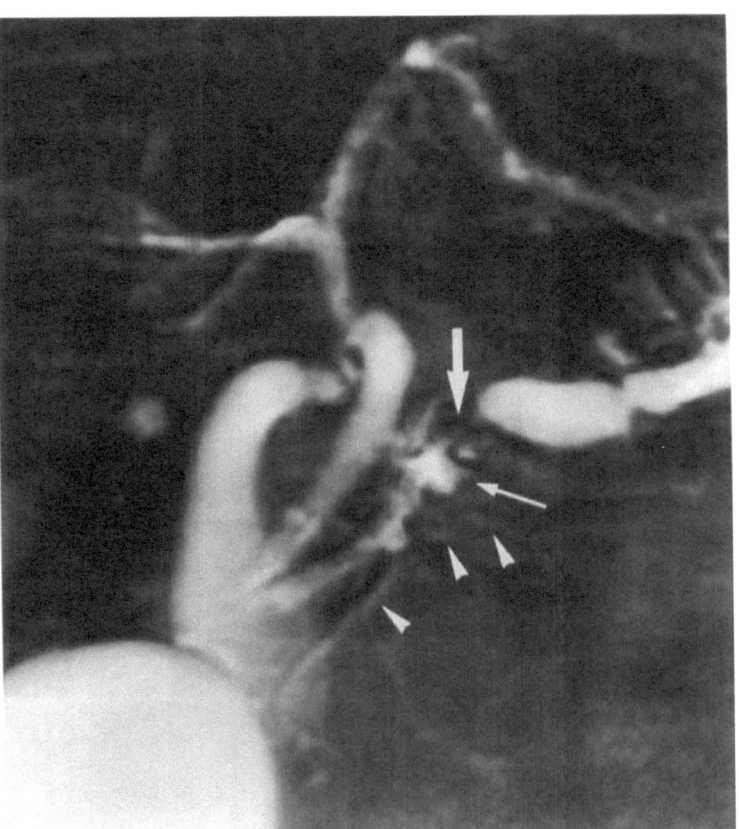

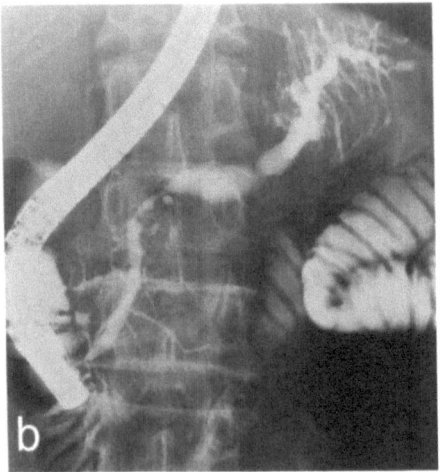

4.3.3   Pancreatic Duct Calculi

Pancreatic duct calculi are pathognomonic for chronic pancreatitis. On MRCP, calculi in the main pancreatic duct are demonstrated as low-signal-intensity foci (signal void or filling defect) surrounded by hyperintense pancreatic juice [3] (Figs. 4-5, 4-8), although protein plugs and neoplasms may show the same findings [16]. The diagnosis of calculi can be confirmed by CT [17].

The reported sensitivity of MRCP in the diagnosis of intraductal filling defect is 92%–100% [3,14]. Calculi in the peripheral branches of the pancreatic ducts or the parenchyma, which are not surrounded by

Fig. 4-8 Pancreatic duct calculi

a. Single slice MRCP shows an intraductal filling defect in the main pancreatic duct (*arrow*) with proximal duct dilatation.

b. Plain film shows a stone in the head of pancreas (*arrow*).

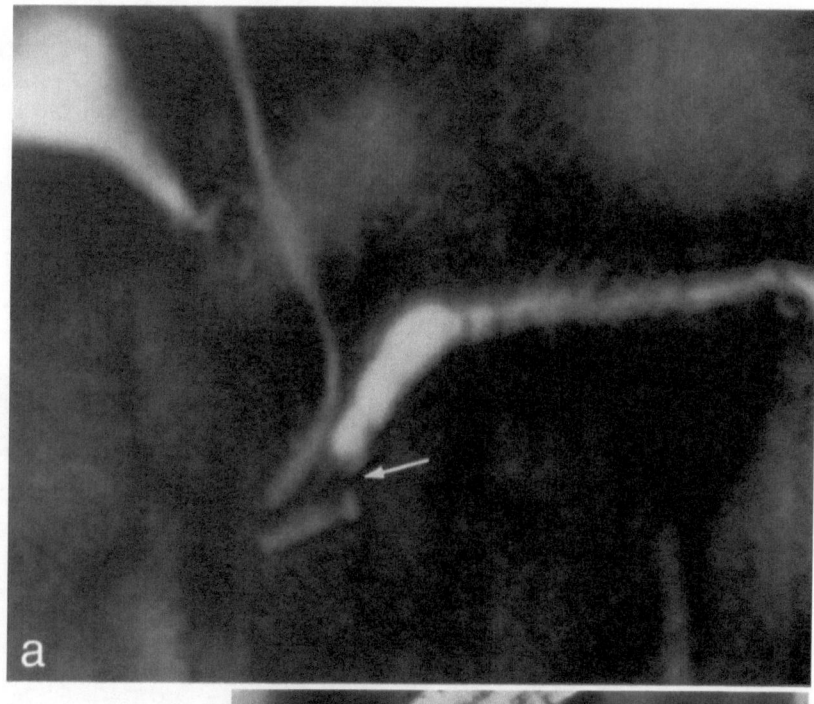

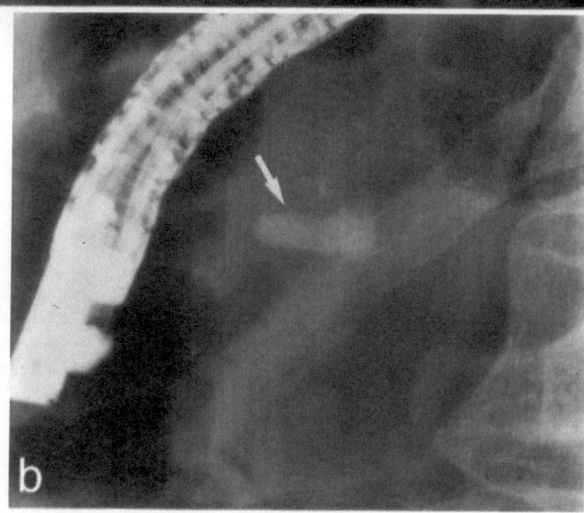

hyperintense pancreatic juice, may not be depicted by MRCP [14].

With an MIP image, it is important to review the source image to confirm the presence of calculi (Fig. 4-9), because stones may be obscured due to the partial volume effect [3].

Fig. 4-9 Pancreatic duct calculi (same cases as in Fig. 4-5)

a. On MIP-MRCP images, pancreatic duct calculi are obscured by the surrounding hyperintense pancreatic juice.
b. By reviewing the source images of MIP, calculi are detected in the main pancreatic duct (*arrows*).

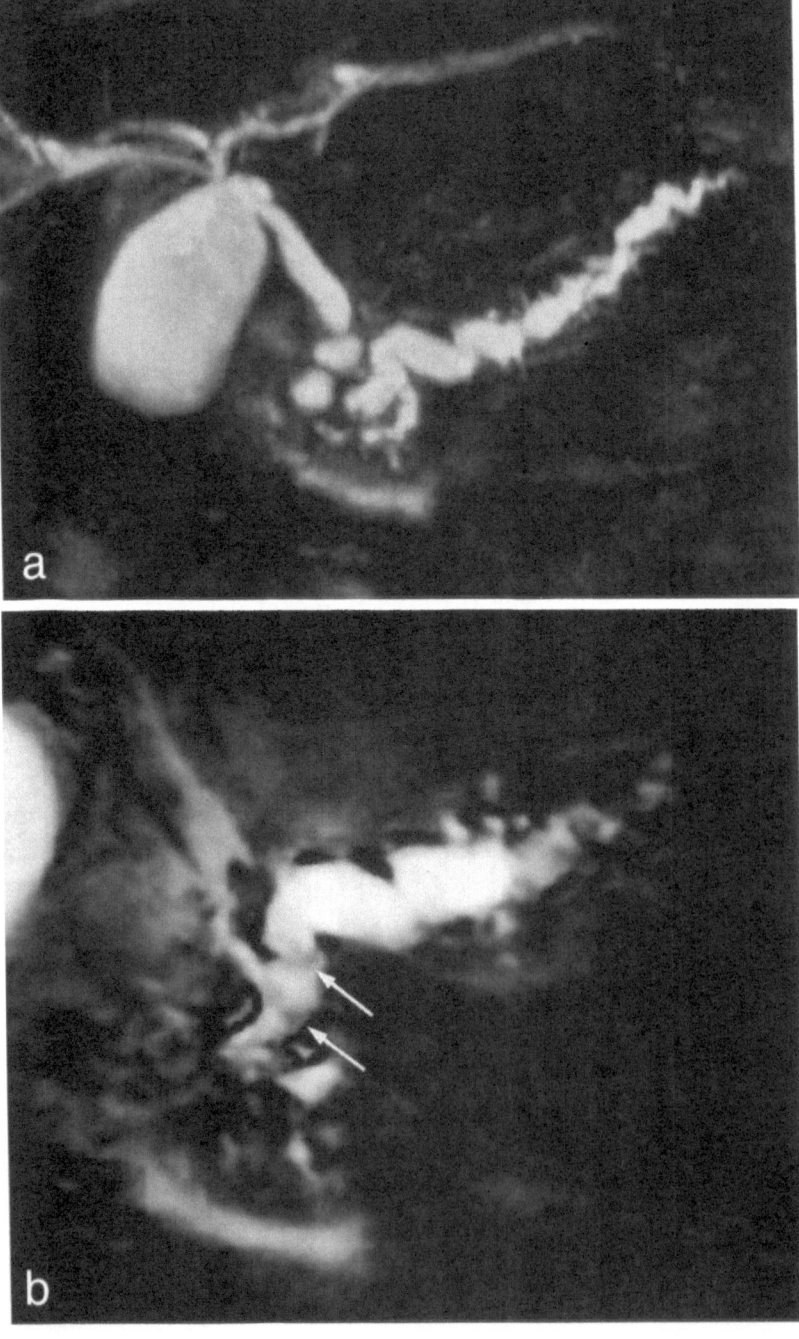

4.3.4   Inflammatory Pancreatic Masses

Inflammatory pancreatic masses are focal enlargements of the pancreatic gland due to acute or chronic inflammation. They are often indistinguishable from pancreatic carcinoma by various diagnostic imaging techniques [18,19], because the pancreatic mass is usually associated with a stricture of the main pancreatic duct or the distal common bile duct (double duct sign [20,21]).

On an ERCP image, a normal pancreatogram, multiple ductal strictures, a long, smooth ductal stenosis, or dilated side branches within the mass indicate inflammatory pancreatic masses [18,21-23]. However, those findings are not specific for inflammatory pancreatic masses and may be observed in pancreatic carcinoma [24].

In our experience of eight patients with inflammatory pancreatic masses, MRCP showed a long, smooth ductal stenosis with dilated side branches within the mass in six patients, and dilated side branches were more clearly depicted on MRCP than on ERCP [25]. Secretin injection improves the image quality and decreases the frequency of false-positive stenosis, because increased output of pancreatic juice distends the main pancreatic duct, resembling the condition in ERCP [9].

Duct penetration within a mass is not specific to inflammatory pancreatic mass, and is sometimes observed in patients with malignant solid tumors [26].

Further diagnostic procedures, such as biopsy, endoscopic ultrasonography, contrast-enhanced sonography, or positron emission tomography, are required to confirm the diagnosis [27-29].

Fig. 4-10 Inflammatory pancreatic mass

a. US shows a mass-like lesion in the head of pancreas (*bold arrow*) and proximal dilatation of the main pancreatic duct (*arrow*).

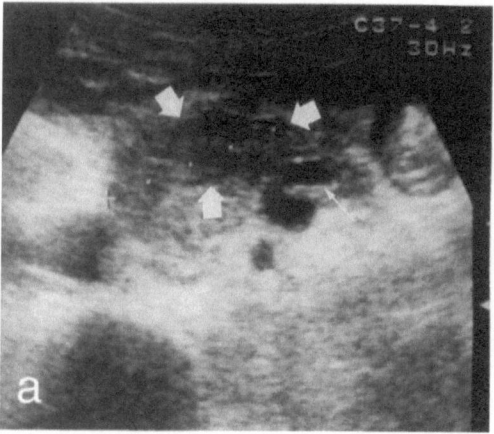

Fig. 4-10 *continued*

b. Single-slice MRCP (before selectin injection) shows stenosis of the main pancreatic duct at the site of the mass (*bold arrow*) with proximal duct dilatation. These findings suggest pancreatic carcinoma.

c. MRCP (after selectin injection) shows a long, smooth stricture in the main pancreatic duct (*arrow*) with isolated dilated side branches corresponding to the site of the tumor, indicating an inflammatory pancreatic mass.

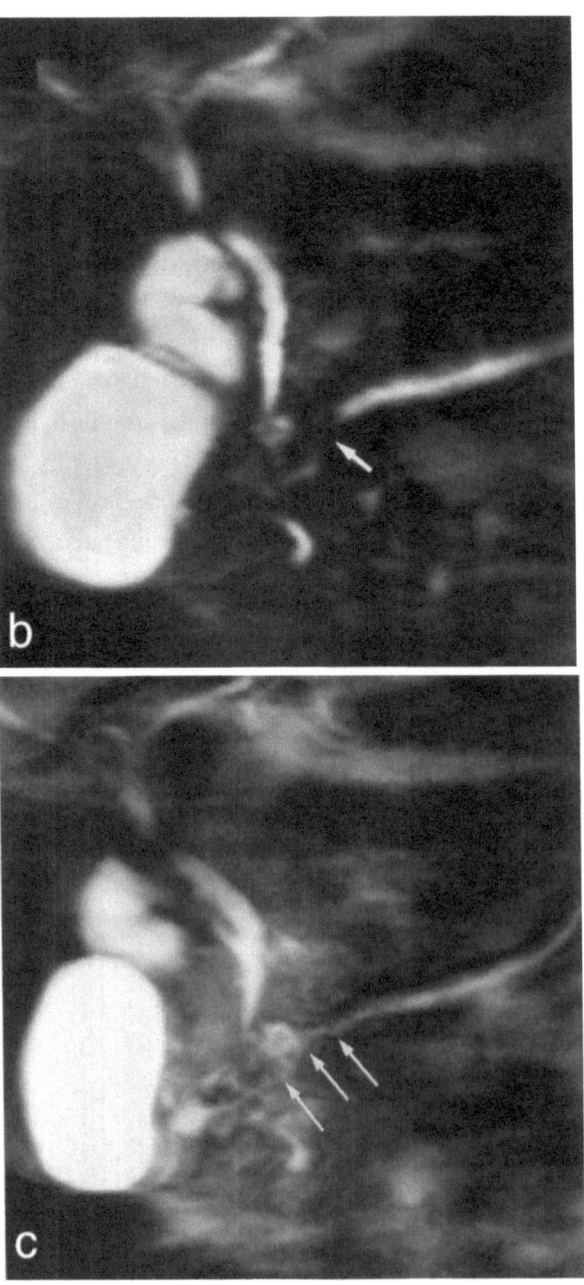

4.3.5   Focal Stenosis of the Main Pancreatic Duct Masquerading as Pancreatic Carcinoma

This condition is a benign fibrous proliferation at the main pancreatic duct and can mimic small pancreatic carcinoma both clinically and radiologically. Pancreatitis is the main cause of the focal stenosis; however, the origin is sometimes unknown [30,31].

MRCP demonstrates a short, smooth stricture of the main pancreatic duct with proximal duct dilatation, and the stricture is different from the abrupt stricture in pancreatic carcinoma. However, biopsy or cytology is required to confirm the diagnosis.

Fig. 4-11 Focal pancreatitis

a. CT shows a low-density area (*arrowheads*) in the body of the pancreas and proximal duct dilatation (*arrow*).

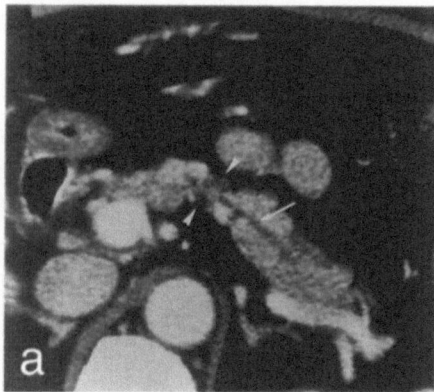

Fig. 4-11 *continued*

b. Single-slice MRCP shows a short stricture of the main pancreatic duct in the body (*arrow*) with proximal duct dilatation. Dilated side branches at the stenosis are observed (*arrowhead*).

c. ERCP shows a short, smooth stenosis (*arrow*) and a dilated side branch. Focal pancreatitis was confirmed histologically.

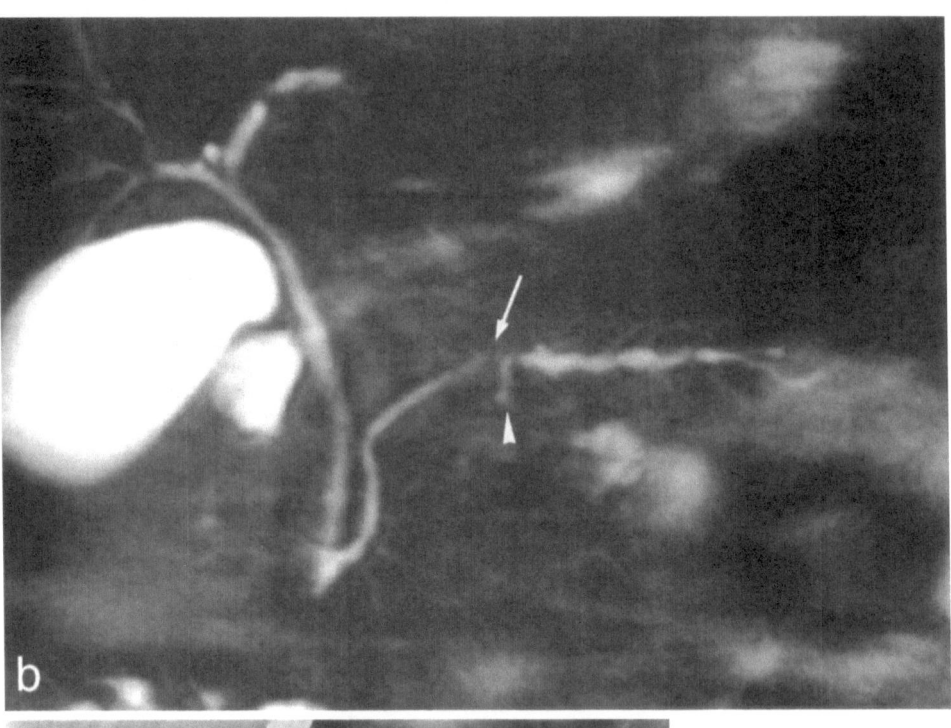

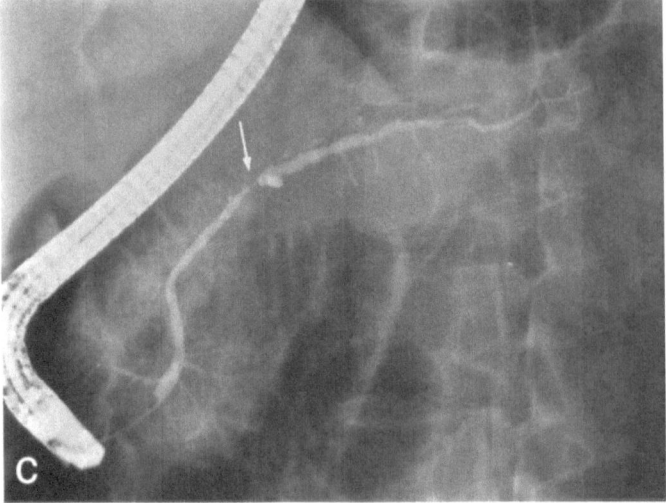

4.3.6 Groove Pancreatitis

Groove pancreatitis is a form of segmental pancreatitis affecting the groove enclosed by the duodenum, the common bile duct, and the head of the pancreas [32,33]. Diagnostic imaging often demonstrates a mass in the head of the pancreas and stenosis of the distal common bile duct and duodenum, so that differentiation between groove pancreatitis and pancreatic carcinoma is often difficult. Pancreatic carcinomas are characterized by abrupt stricture of the main pancreatic duct and the common bile duct. In contrast, groove pancreatitis usually shows slight changes in the pancreatic duct and a long, smooth stenosis of the distal common bile duct [34]. Also, cysts, either true cysts or pseudocysts, are seen in the groove or the duodenal wall.

MRCP can depict the changes in both the pancreatic duct and the common bile duct simultaneously, and demonstrates cysts in the groove or the duodenal wall [35]. Therefore, MRCP is useful for the diagnosis of groove pancreatitis.

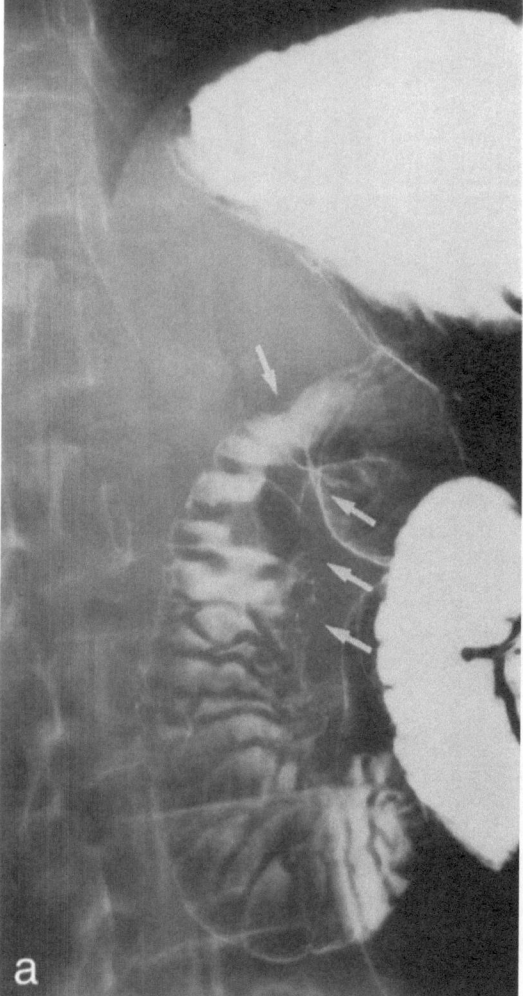

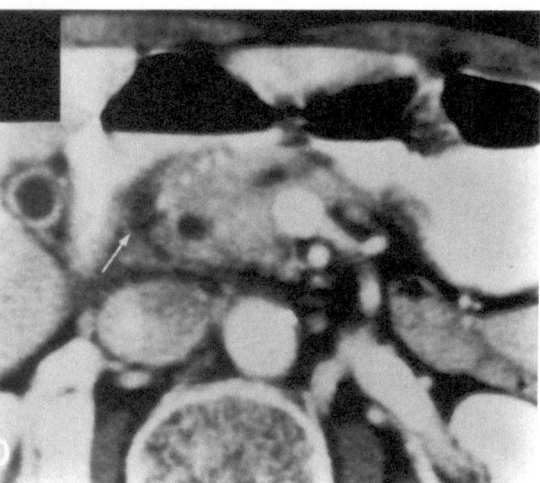

Fig. 4-12 Groove pancreatitis

a. Hypotonic duodenography shows irregular stenosis in the second portion of the duodenum (*bold arrows*).
b. CT shows a low-density area (*arrow*) in the groove enclosed by the enlarged pancreas head, the duodenum, and the common bile duct.

Fig. 4-12 *continued*

c. Single-slice MRCP shows a long, smooth stenosis of the distal common bile duct (*arrows*) and slight dilatation of the main pancreatic duct (*arrowhead*). A cyst in the groove is also shown.

d. ERCP shows findings similar to those in MRCP.

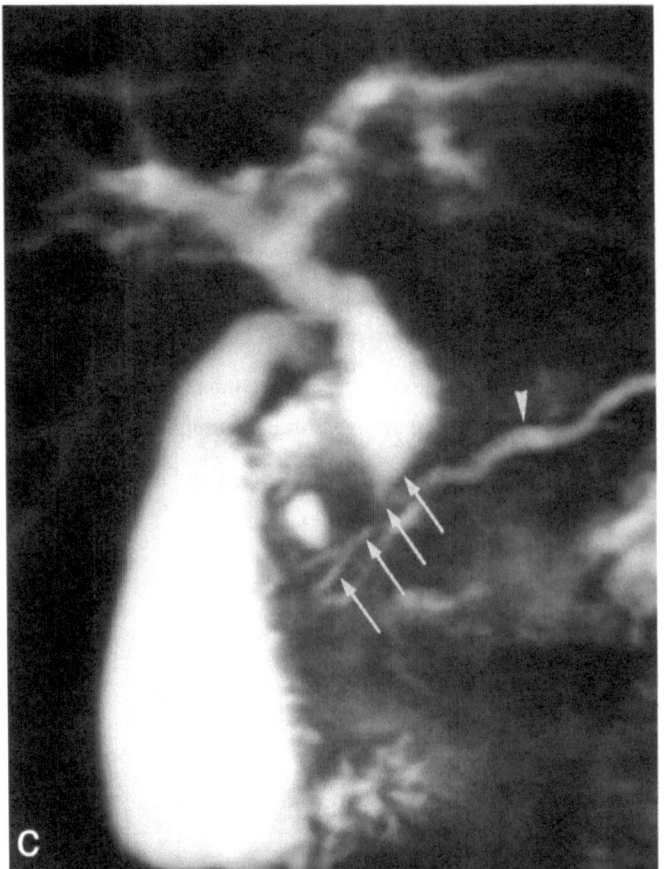

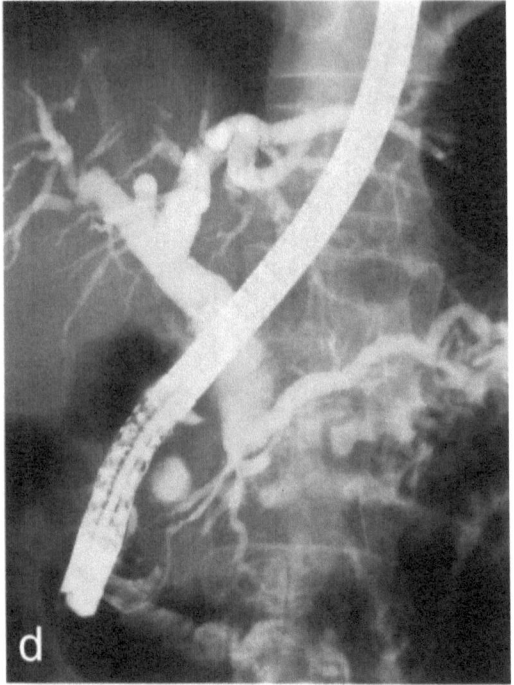

4.4   Cystic Lesions of the Pancreas

MRCP has high sensitivity for the detection of cystic lesions of the pancreas [36]. In our study of 75 patients with pancreatic cysts, the sensitivity of MRCP, US, and CT was 98%, 70%, and 87%, respectively.

MRCP delineates detailed morphology and precise location of the cyst, and simultaneously its relationship with the main pancreatic duct. Furthermore, 3D-MRCP with MIP can depict the internal structure of the cyst, including the septa and papillary protrusion, by reviewing the source images [36-39] (Figs. 4-16e, 4-19c).

However, MRCP cannot depict the presence of mucin and cannot directly confirm the communication between the cyst and the main pancreatic duct, which is possible only through injection of contrast material as in ERCP [36-39].

4.4.1   Pancreatic Pseudocysts

Pseudocysts of the pancreas are round fluid masses and most commonly develop secondary to acute or chronic pancreatitis. The size of the cyst may increase or decrease according to the inflammatory activity. Asymptomatic pancreatic pseudocysts often disappear spontaneously. However, patients with symptomatic pseudocysts require close observation and sometimes percutaneous drainage, because the pseudocysts may be associated with severe complications such as bleeding, rupture, and infection [40,41].

On MRCP, pseudocysts of the pancreas are depicted as round lesions of high signal intensity in the pancreas or in the peripancreatic spaces [42]. MRCP not only depicts detailed morphology and precise location of the cysts including unusual location adjacent to the spleen, liver, and mediastinum [43], but also demonstrates stenosis and dilatation of the pancreatic duct, calculi in the pancreatic duct, and peripancreatic fluid collection (see section 4.3.1). Therefore, correct diagnosis of the underlying pathology such as chronic pancreatitis or acute pancreatitis is feasible [44-46]. The relationship between a pseudocyst and the surrounding structures such as the biliary tract and stomach is also revealed with MRCP. In patients with a past episode of intracystic hemorrhage, the signal intensity of the cystic content is markedly reduced due to the blood degeneration (Fig. 4-23).

MRCP can be conducted conveniently and is therefore useful in following the clinical courses of patients with pseudocysts [8].

Fig. 4-13 Pancreatic pseudocyst

a. CT shows atrophy of the pancreatic parenchyma, dilatation of the main pancreatic duct (*arrow*), and pseudocyst in the tail of the pancreas (*bold arrows*).

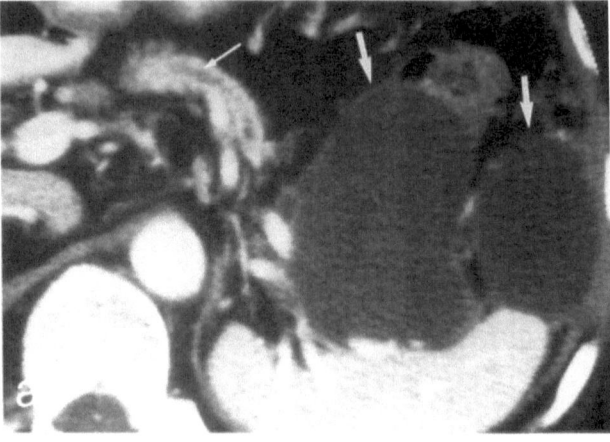

Fig. 4-13 *continued*

b. Single-slice MRCP shows two cystic lesions (*bold arrows*) in the tail of the pancreas and a stenosis of the main pancreatic duct at the distal portion of the cysts (*arrow*) with proximal dilatation (*arrowhead*). Stenosis is also found in the head of the pancreas (*open arrow*). No communication is seen between the two cysts, and percutaneous drainage is required in both cysts.

c. ERCP shows only pancreatic duct obstruction in the head of the pancreas (*arrow*).

d. In percutaneous drainage and cystography, the cyst was incompletely visualized, and further drainage is required as shown in MRCP.

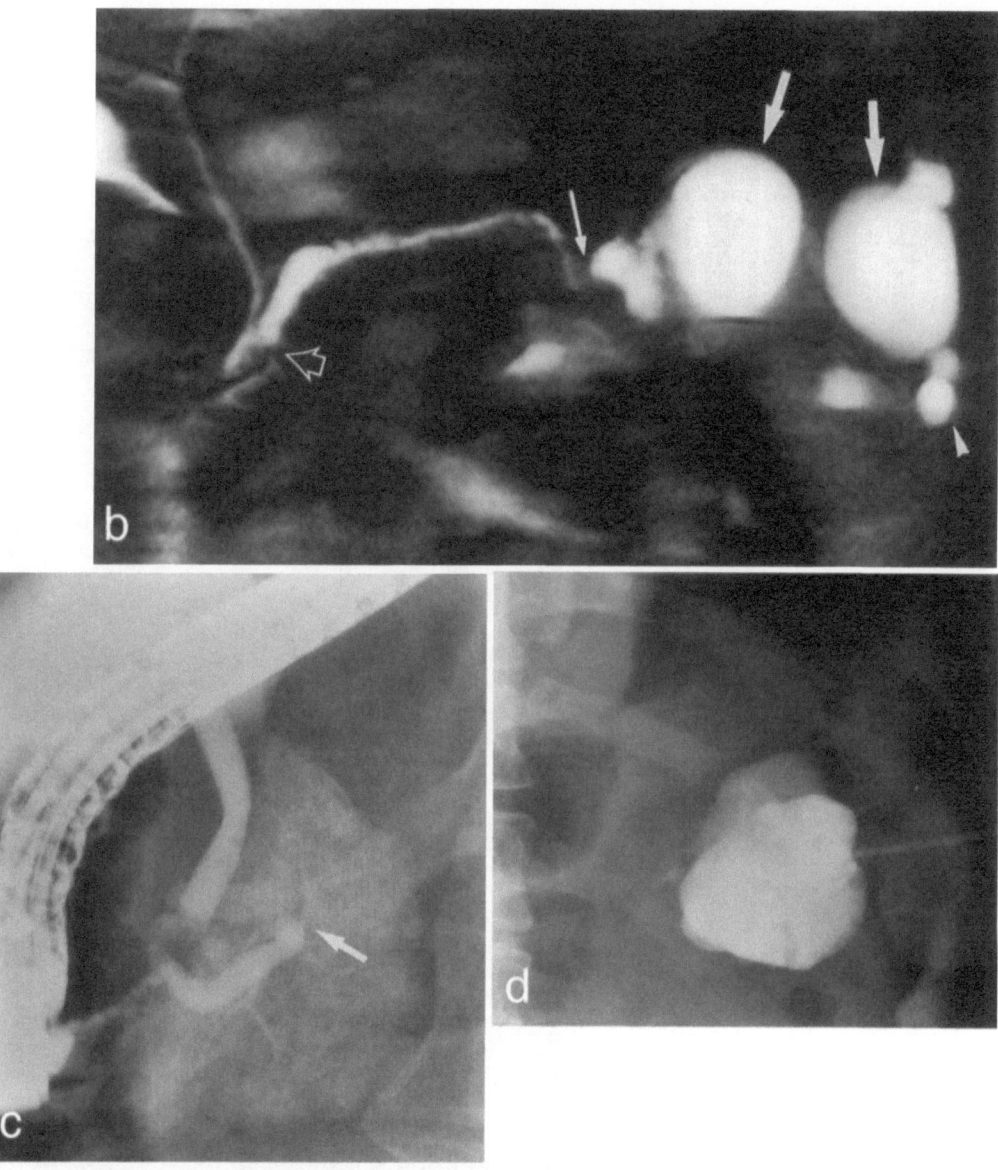

4.4.2   Solitary True Cyst

Solitary true cysts are rare among the cysts of the pancreas. These cysts are non-neoplastic, unilocular, lined with a single layer of cuboidal or flat epithelium, and rarely symptomatic. Most of the solitary true cysts are found incidentally, and their etiology is still unknown [47-49].

MRCP demonstrates round, unilocular cysts without communication with the main pancreatic duct. MRCP is useful in follow-up study of the solitary true cyst of the pancreas.

Fig. 4-14 Solitary true cyst

Single-slice MRCP shows a round unilocular cyst (*arrow*) in the body of the pancreas.

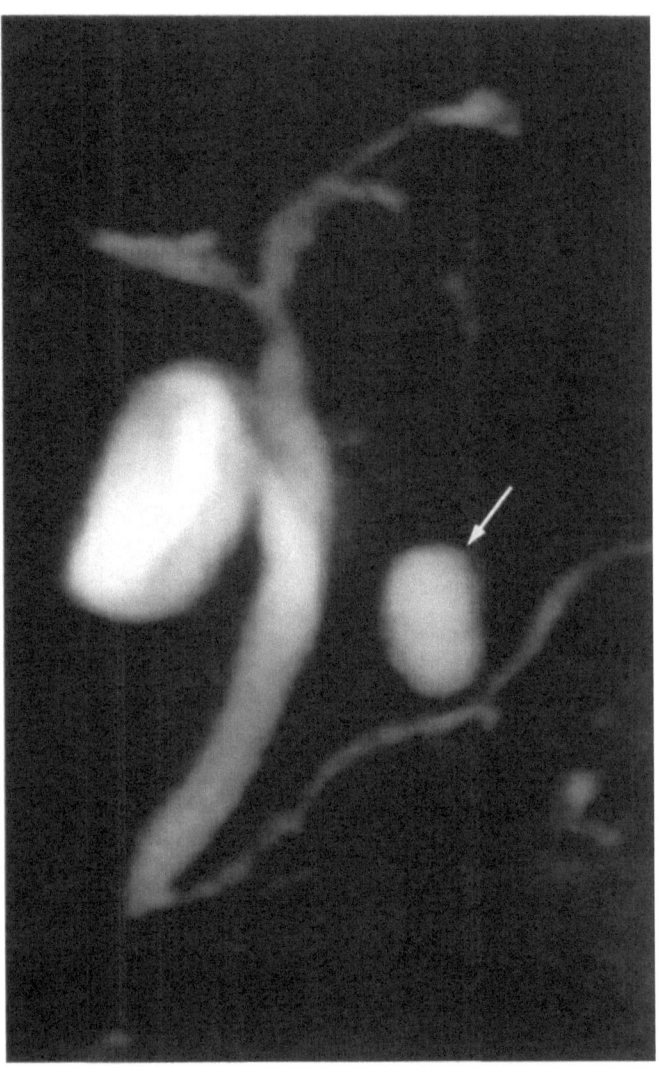

4.4.3   Intraductal Papillary Mucinous Tumors

Intraductal Papillary Mucinous Tumors (IPMTs) [50] include hyperplasias, adenomas, and adenocarcinomas [51]. IPMTs are lined by tall columnar epithelium with mucin hypersecretion. Papillary proliferation of the epithelium is common, showing mild, moderate, or severe cellular atypia. Due to hypersecretion of mucin, the pancreatic duct system is dilated [52].

Resection may not be mandatory in hyperplasia, but adenomas and carcinomas should be subject to surgical intervention. Diameter of the main pancreatic duct greater than 5 mm, dilatation of the side branch larger than 3 cm, nodular excrescences in the duct system, and papillary proliferation of the duct epithelium greater than 2 mm indicate intraductal papillary adenocarcinoma [53,54] (Fig.4-15).

At ERCP examination, mucin secretion from the orifice of the papilla of Vater is observed, and this finding confirms the presence of IPMT [50]. However, the pancreatic duct system is difficult to opacify with contrast medium injection due to the presence of mucin, and differentiation of nodular excrescences from mucin is often difficult.

MRCP allows complete visualization of the dilated pancreatic duct system regardless of mucin (Fig.4-16B) [36-39,55]. On MRCP, the signal intensity of mucin appears to be the same as that of pancreatic juice. On 3D-MRCP images, the septa (Fig. 4-16d), nodular excrescences (Figs. 4-16e, 4-19d), and the communication between the dilated side branches and the main pancreatic duct are depicted by reviewing the source images (Fig. 4-16c,f) [36-39,55]. Mural nodules and mucin are readily differentiated.

MRCP is useful for follow-up study of IPMT (Fig. 4-18, 4-21).

Fig. 4-15 Findings indicating intraductal papillary adenocarcinomas

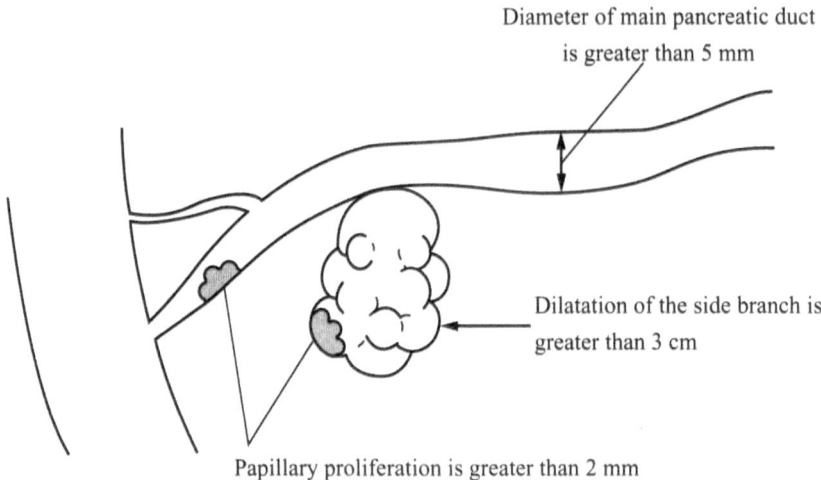

Diameter of main pancreatic duct
is greater than 5 mm

Dilatation of the side branch is
greater than 3 cm

Papillary proliferation is greater than 2 mm

Fig. 4-16 Intraductal papillary adenoma

a. ERCP shows the dilated side branch (*arrow*) inadequately opacified with contrast medium due to hypersecretion
of mucin.

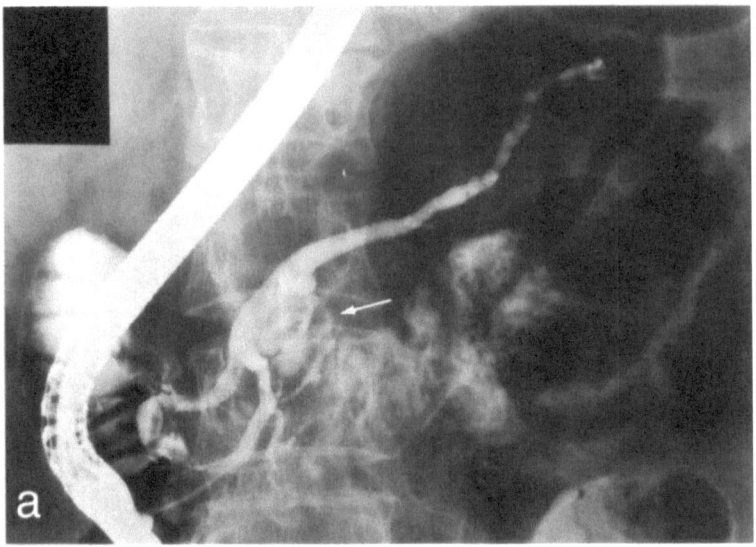

Fig. 4-16 *continued*

b. Three-dimensional MRCP with MIP allows complete visualization of the dilated side branches (*arrows*). Multiple dilated side branches are depicted but are not demonstrated by ERCP.

c. By rotating 3-D images along the main pancreatic duct, the relationship between the dilated side branch (*arrow*) and the main pancreatic duct (*arrowhead*) is revealed.

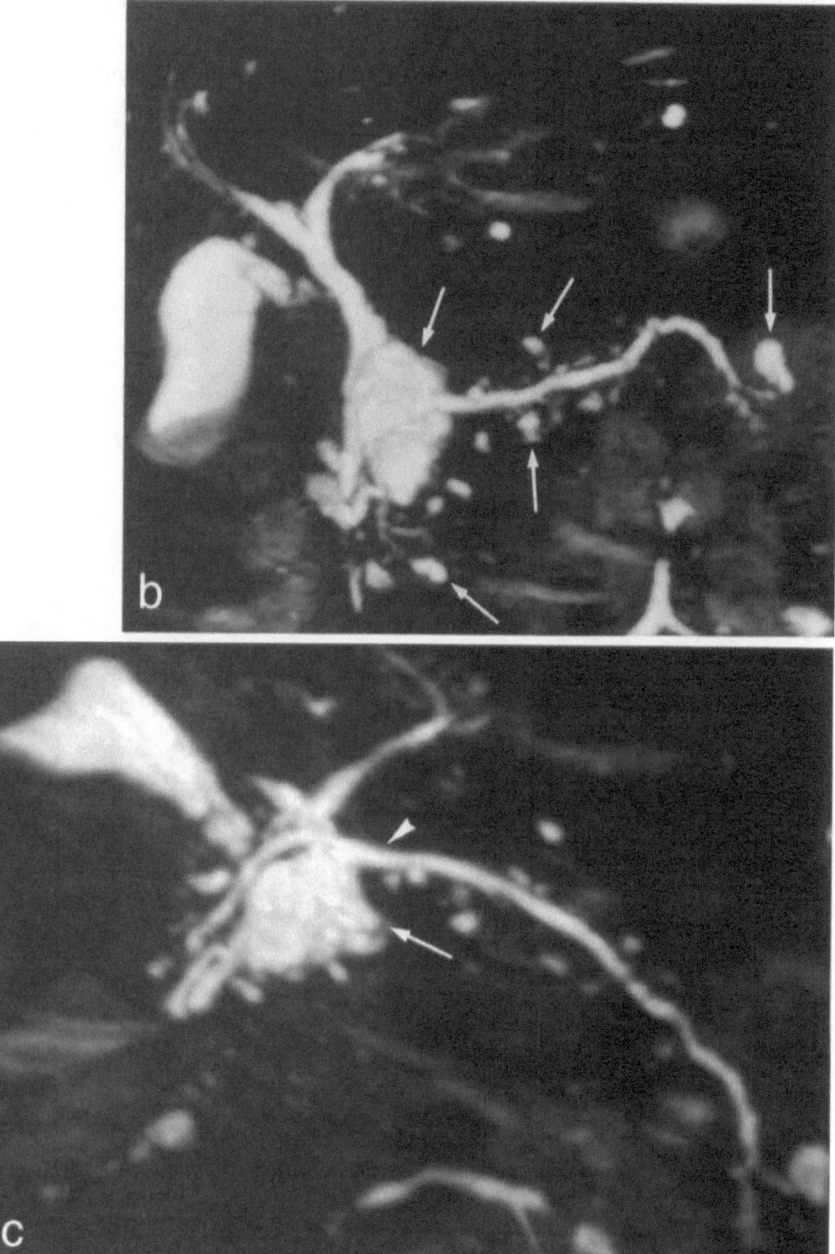

Fig. 4-16 *continued*

d. The source image of MIP shows the septa (*arrowheads*) in the dilated side branch adjacent to the common bile duct (*arrow*).

e. By reviewing the source images at 1-mm intervals, a papillary protrusion (*arrow*) in the dilated side branch is demonstrated.

f. Another source image shows communication (*arrow*) between the dilated side branch (*bold arrow*) and the main pancreatic duct (*arrowheads*).

g. Histology shows papillary protrusion in the dilated side branch, with a diagnosis of papillary adenoma.

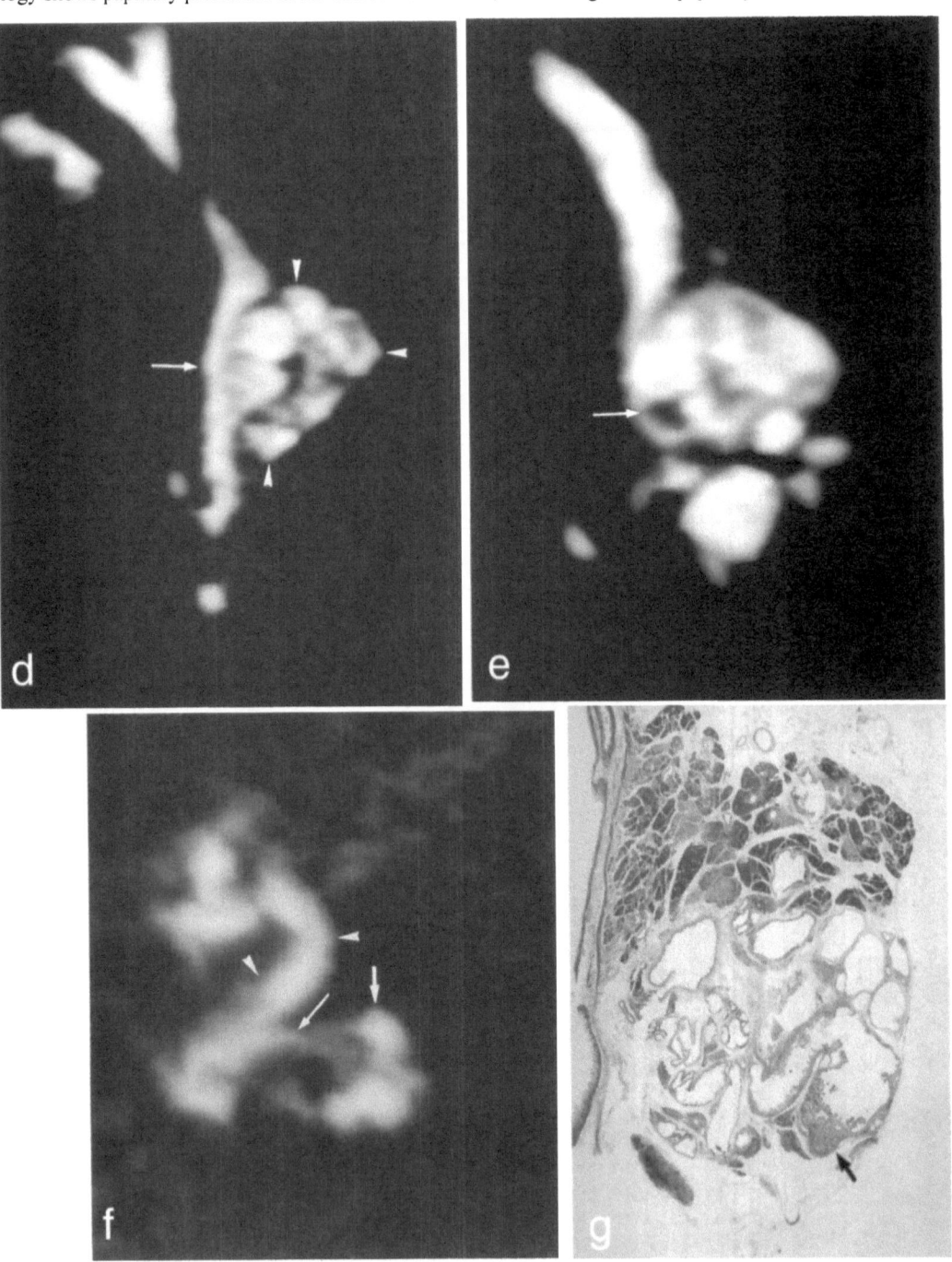

Fig. 4-17 Intraductal papillary adenoma of the main ductal type

a. On ERCP, due to hypersecretion of mucin, the pancreatic duct system is not well visualized.

b. Single-slice MRCP shows the entire pancreatic duct system (*arrow*).

c. Endoscopic US shows elevated lesions (*arrowheads*) in the dilated main pancreatic duct (*arrow*). Histology verified papillary adenoma in the main pancreatic duct.

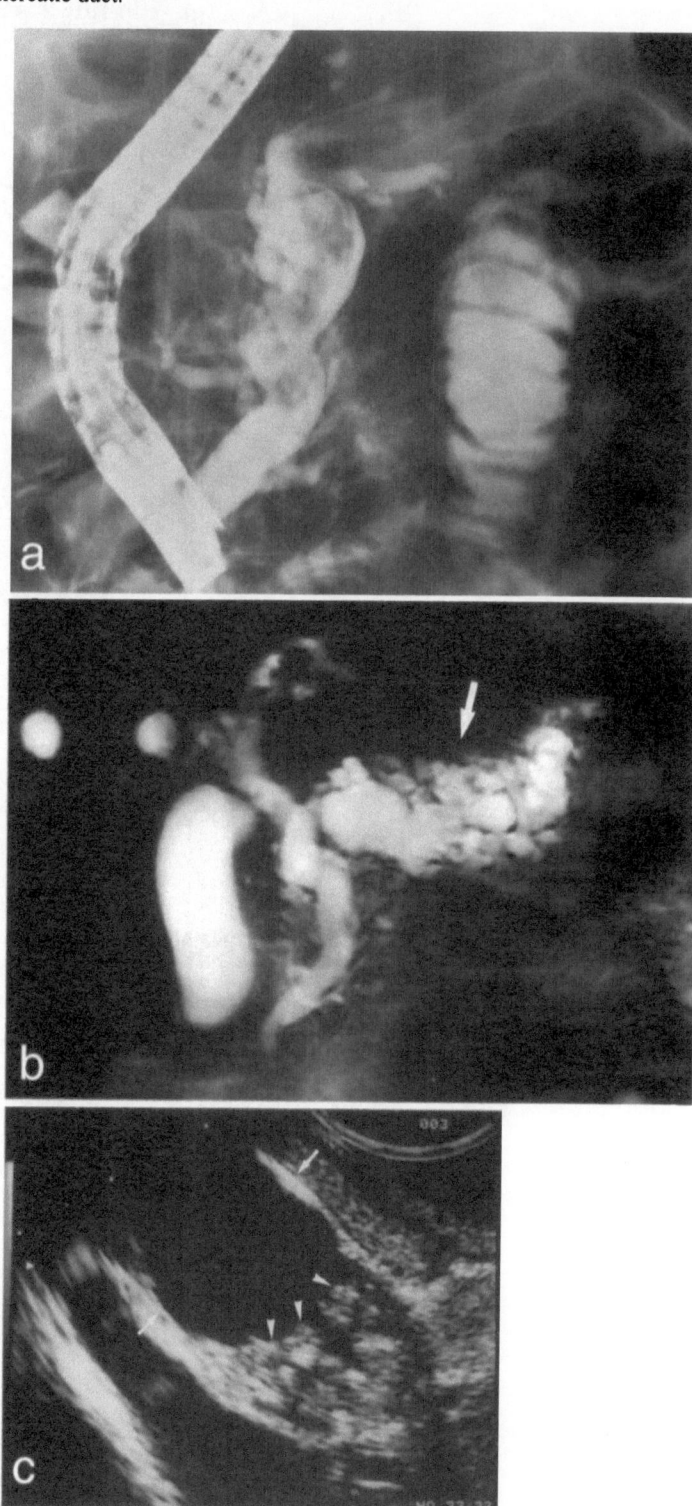

Fig. 4-18 Intraductal papillary adenoma (under observation)

a. MRCP shows cystic dilation of the main pancreatic duct in the tail.

b. Follow-up MRCP 4 months after initial observation shows that the dilated main pancreatic duct (*arrow*) has increased in diameter. Histology revealed papillary adenoma in the main pancreatic duct.

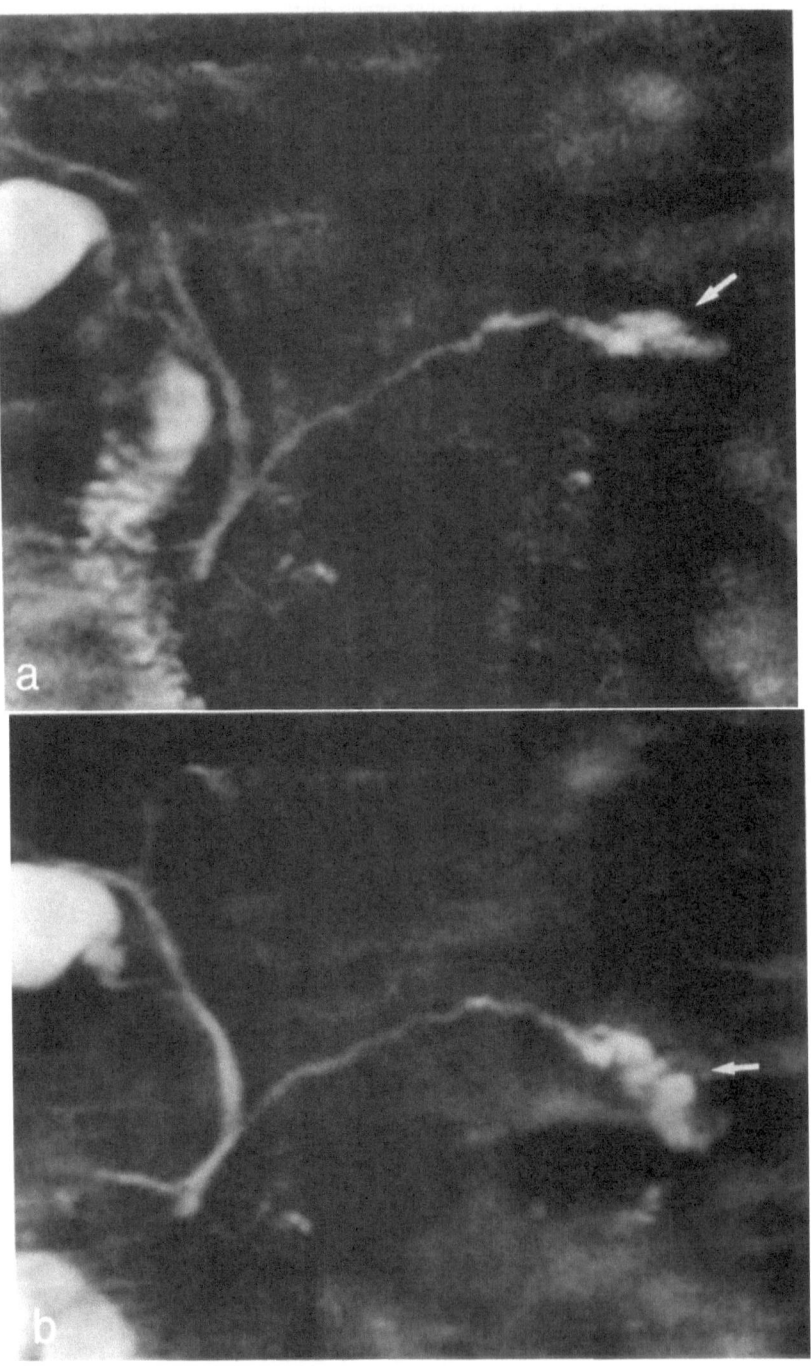

Fig. 4-19 Intraductal papillary adenocarcinoma

a. ERCP shows dilation of the main pancreatic duct, but fails to depict the side branch.

b. Three-dimensional MRCP with MIP clearly shows the entire dilated side branch (*arrow*) together with the main pancreatic duct.

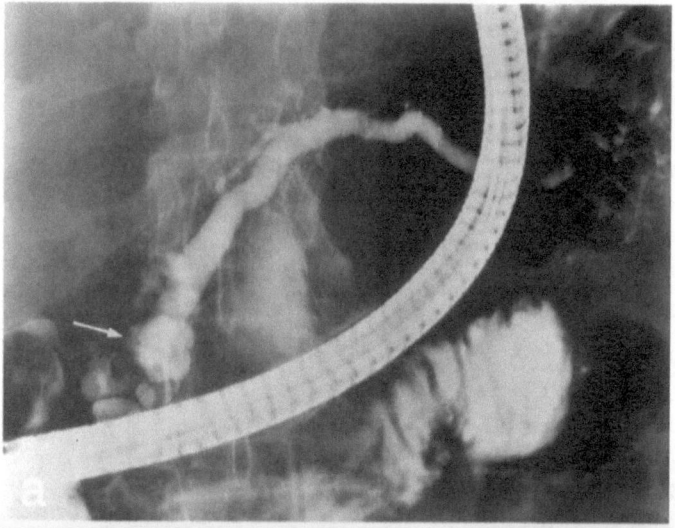

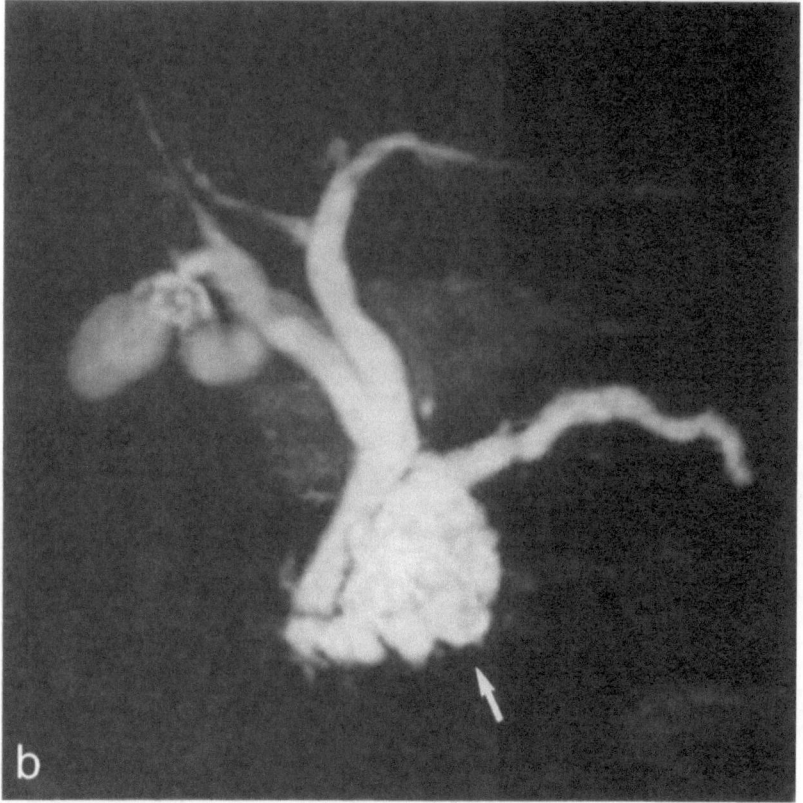

Fig. 4-19 *continued*

c. By reviewing the source images at 1-mm intervals, a papillary protrusion (*arrow*) in the side branch is demonstrated.

d. Endoscopic US (linear type) confirms papillary projections (*arrowheads*) as in MRCP.

e. Histology shows papillary projection in the ectatic side branch, and is diagnosed as adenocarcinoma.

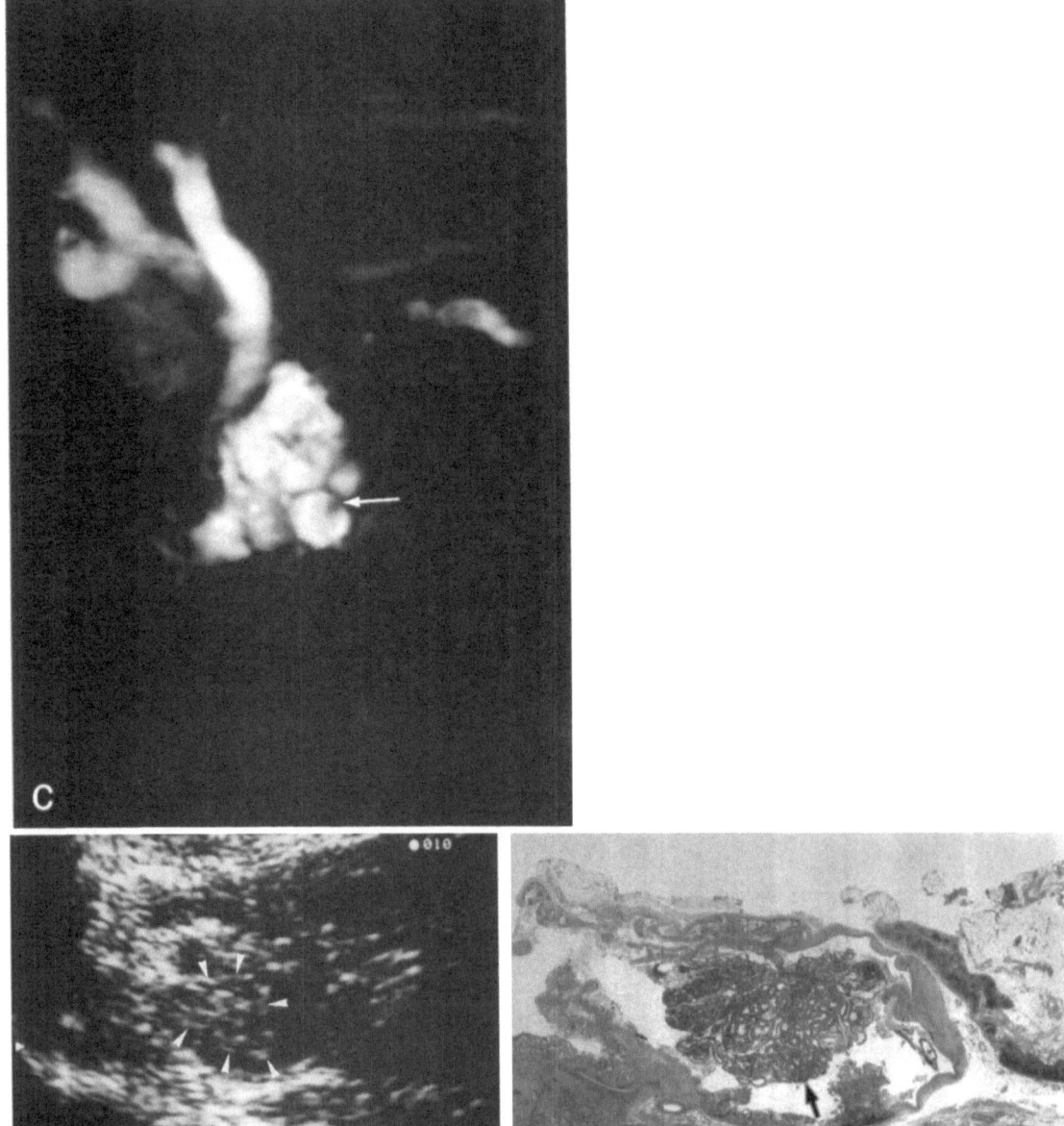

**Fig. 4-20 Intraductal papillary
adenocarcinoma with
parenchymal invasion**

a. MRCP shows dilatation of the
main duct (*arrow*) and an irregular
high-signal intensity area in the
parenchyma in the head of the
pancreas (*arrowheads*).
b. ERCP shows dilatation of the
main pancreatic duct (*arrow*) and a
defect in the side branches of the
pancreas (*arrowhead*).
c. Histology shows intraductal
papillary adenocarcinoma with
parenchymal invasion.

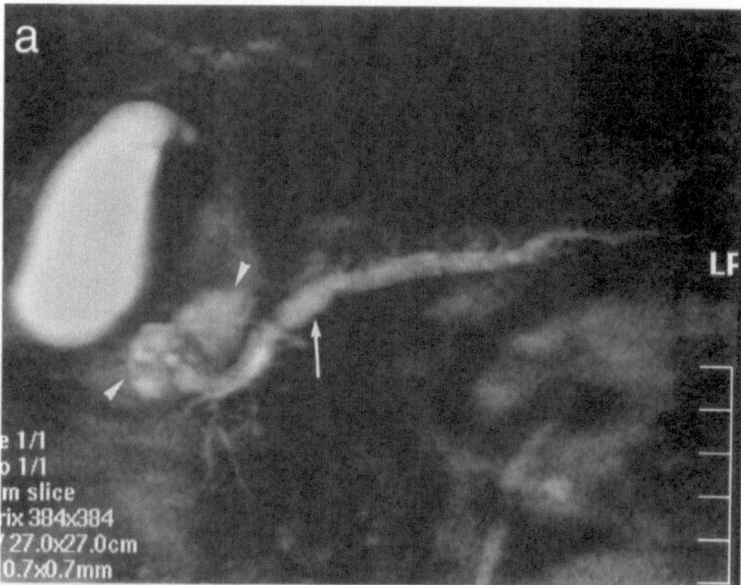

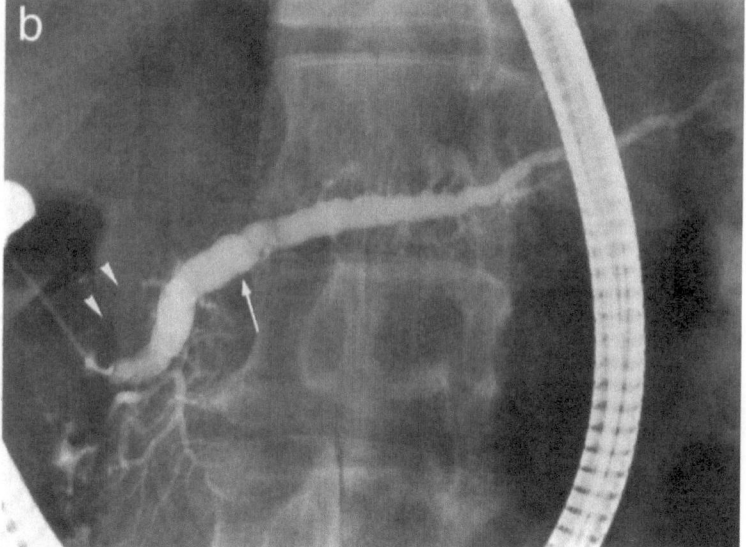

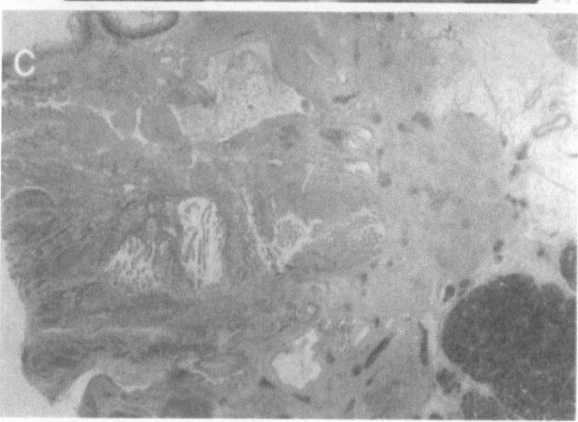

Fig. 4-21 Intraductal papillary hyperplasia (follow-up)

a. Single-slice MRCP shows ectatic changes of the main pancreatic duct and branches (*arrow*).

b. Follow-up MRCP 2 years and 7 months later shows no changes compared with the initial MRCP.

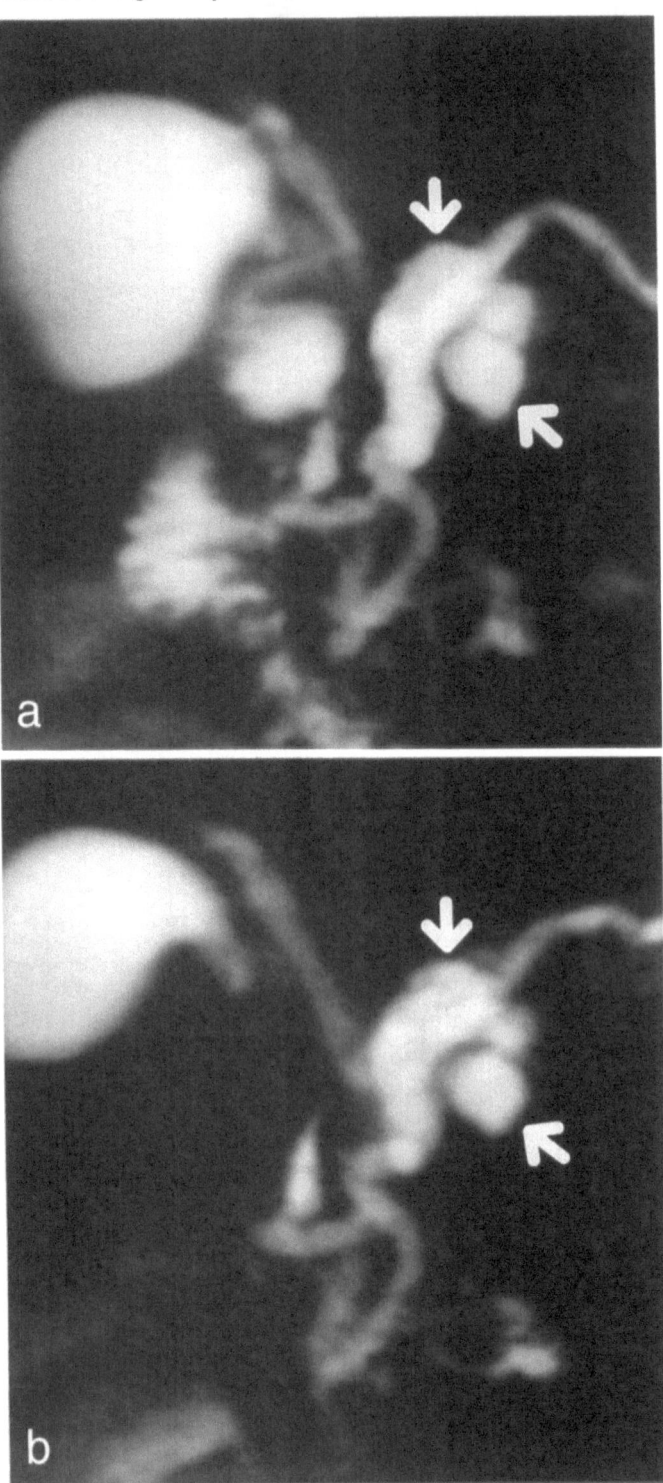

4.4.4   Mucinous Cystic Neoplasm of the Pancreas

Mucinous cystic neoplasms (mucinous cystadenoma, cystadenocarcinoma) occur commonly in middle-aged women and are usually located in the body or tail of the pancreas. They are well encapsulated with a thick fibrous wall and a multilocular internal structure. The cysts are lined by mucinous or columnar epithelium with papillary proliferation, which shows mild, moderate, or severe cellular atypia representing juxta-position of benign and malignant epithelium. The fluid within the cysts is mucinous or hemorrhagic. Indications of malignancy include irregular outline of the cyst wall with tissue projecting into the lumen. Other signs of malignancy are invasion of adjacent organs and metastases to the liver and lymph nodes [44,56-59].

ERCP shows focal irregular narrowing, occlusion, or displacement of the main pancreatic duct at the corresponding location of the cyst.

MRCP provides entire images of the main pancreatic duct and the cyst simultaneously. Rotating the MIP image around the main pancreatic duct, the relationship between the cyst and the main pancreatic duct is clearly demonstrated (Fig. 4-22c). The septa and mural nodules in the cyst are visualized by reviewing the source images (Fig. 4-22d). In addition, the differences in mucus composition in individual lobules separated by septa are shown as differences in signal intensity [36,60]. Marked reduction in signal intensity of the cyst indicates intracystic hemorrhage due to the blood degeneration product (Fig. 4-23).

Fig. 4-22 Mucinous cystadenoma

a. ERCP shows displacement of the main pancreatic duct at the corresponding location of the cyst (*arrow*).

b. Three-dimensional MRCP with MIP provides entire images of the main pancreatic duct (*arrowhead*) and the cyst (*bold arrow*) simultaneously. An area of strong signal intensity is seen in the cyst (*arrow*), suggesting a lobule separated by septa.

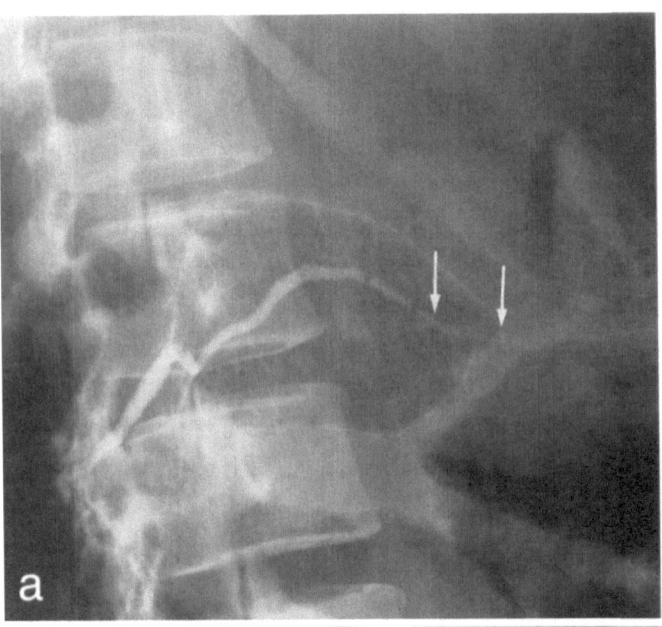

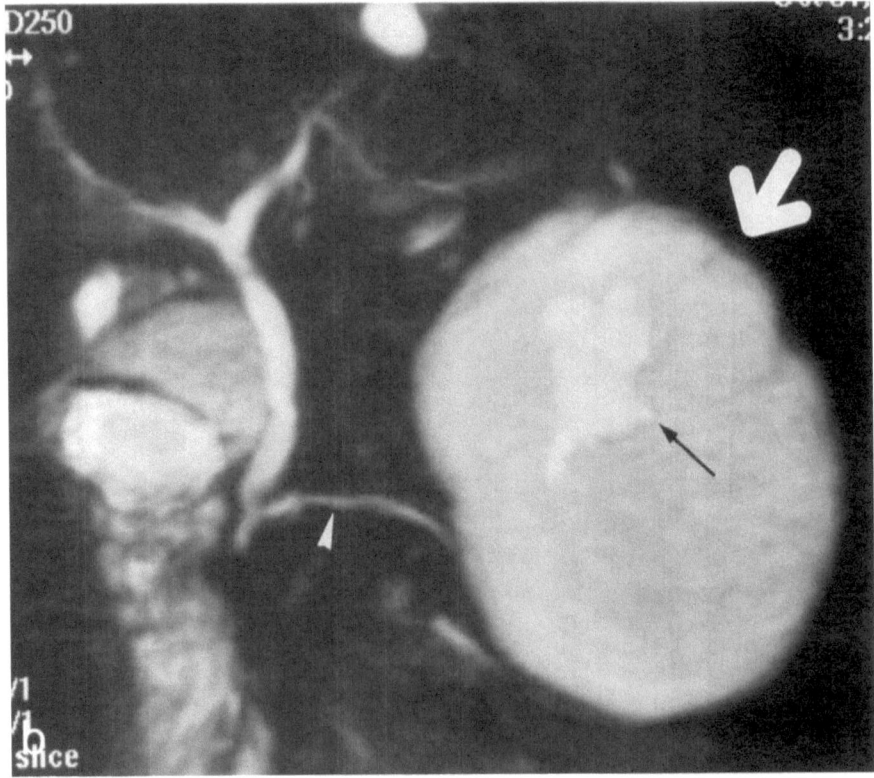

Fig. 4-22 *continued*

c. By rotating the 3-D images around the main pancreatic duct, the relationship between the cyst (*bold arrow*) and the main pancreatic duct (*arrowhead*) is clearly demonstrated.

d. The source image shows septa in the cyst (*arrows*). Differences in mucus composition within each lobule separated by septa are shown as differences in signal intensity.

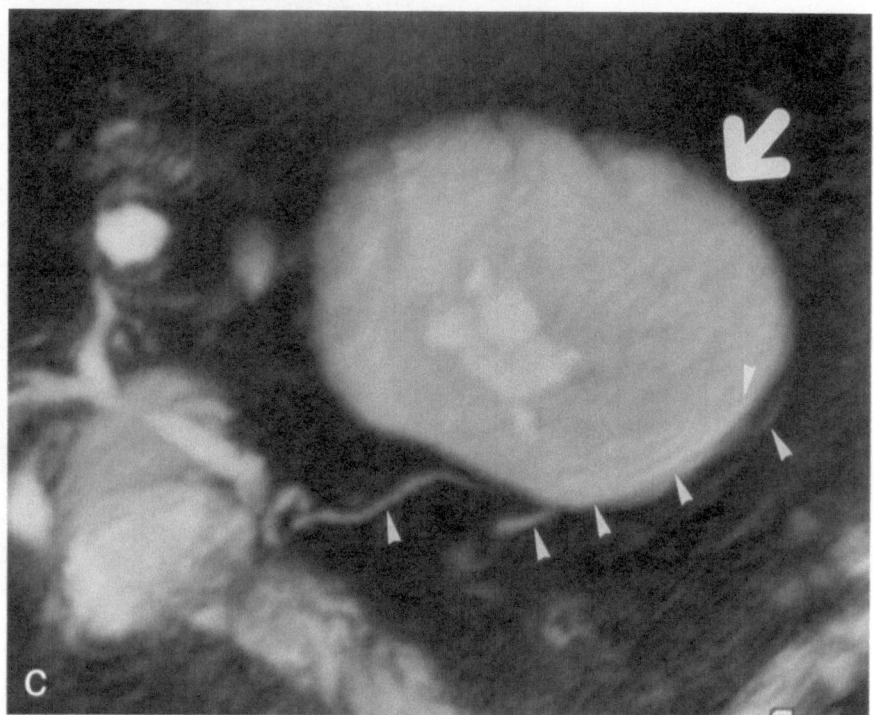

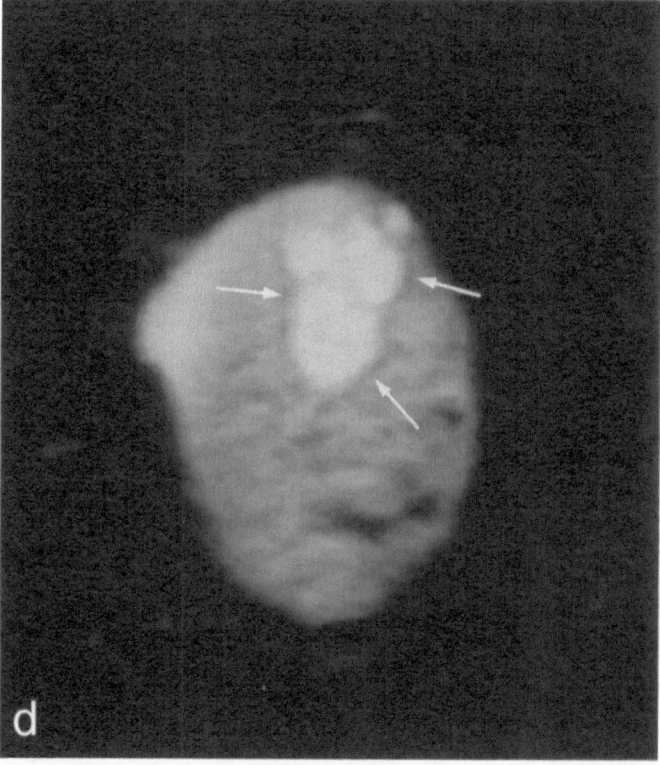

Fig. 4-23 Mucinous cystadenoma with intracystic hemorrhage

Marked reduction in signal intensity of the cyst indicates intracystic hemorrhage.

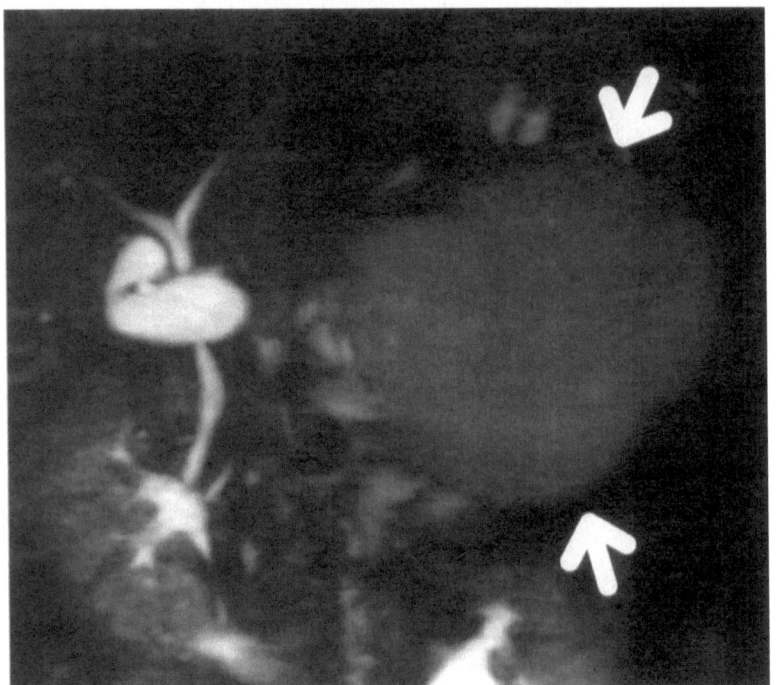

4.4.5   Serous Cystadenoma of the Pancreas

Serous cystadenomas occur most frequently in middle-aged women [57,61-63] and are located in the body or tail of the pancreas [64,65]. The tumors are composed of numerous small cysts, varying in size from a few millimeters to several centimeters, that frequently have a stellate, fibrotic scar or calcification at the center. The cysts contain clear, serous fluid. The cysts are lined by a single layer of epithelium in which the cytoplasm is rich in glycogen [57,61-63]. The main pancreatic duct is normal in 50% of cases, and the cysts do not communicate with the duct system [66].

On MRCP, serous cystadenomas appear as multilocular high signal-intensity-masses [36,60]. The change of the main pancreatic duct and the relationship between the cyst and the duct system is depicted simultaneously.

Fig. 4-24 Serous cystadenoma

a. Contrast-enhanced CT shows cystic lesion with honeycomb appearance (*arrowheads*) in the body of the pancreas.
b. MRCP shows multilocular high signal-intensity-mass (*arrow*) in the body of the pancreas, a typical appearance of serous cystadenoma.

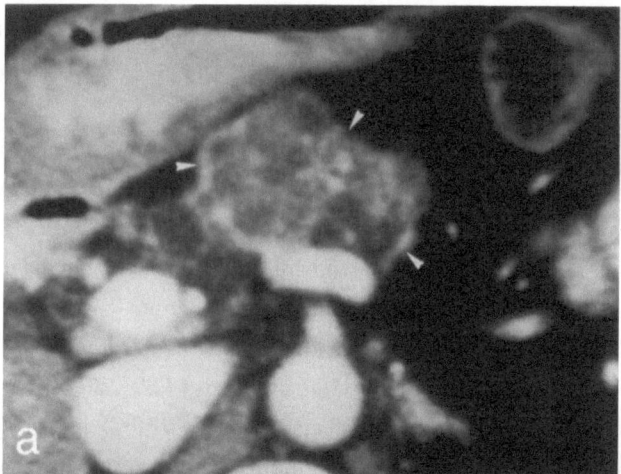
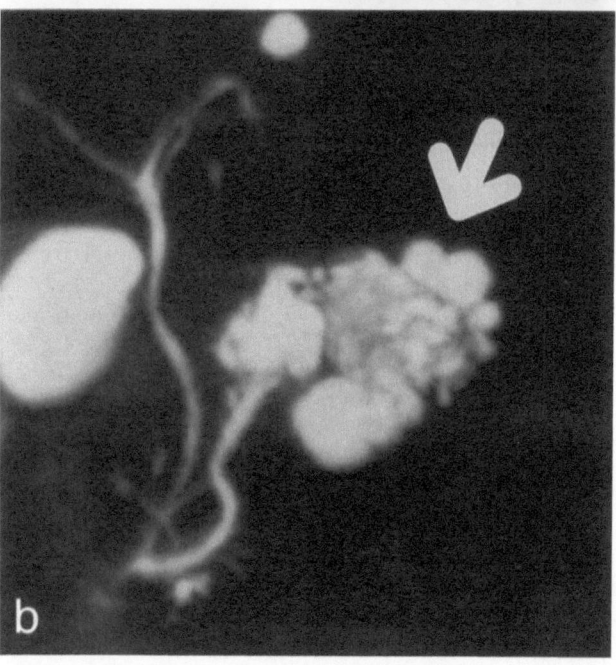

4.5 Pancreatic Ductal Adenocarcinomas

Pancreatic ductal adenocarcinomas have had a steadily increasing impact on most populations throughout the world as the incidence of these carcinomas has risen. Pancreatic carcinomas are the fifth leading cause of death from cancer in Japan. Five-year survival rates by Japan Pancreas Society stage were 46.3% in stage I, 27.5% in stage II, 20.4% in stage III, and 8.3% in stage IV, indicating the significantly better prognosis in early-stage pancreatic carcinoma. Therefore, it is important to make any effort to diagnose early pancreatic carcinomas [67-76].

ERCP has been found to be highly accurate in the diagnosis of pancreatic carcinomas. Accuracy rates of between 90% and 100% have been reported in the literature. Abnormalities of the pancreatic duct system in pancreatic carcinomas include obstruction, stenosis with proximal duct dilatation, extraductal extravasation into the tumor or duct remnant, and filling defect in the dilated duct [20,21]. Because pancreatic ductal adenocarcinomas arise from the duct epithelium, ERCP can demonstrate very small lesions. Some of the ductal changes seen in ERCP are not specific to carcinoma, and require further investigation or other supportive evidence to establish the diagnosis [70,73,75-77].

MRCP shows the same ductal changes as ERCP in pancreatic carcinomas. MRCP findings of 60 cases of pancreatic ductal adenocarcinoma examined in our department have been summarized [75]. Of the pancreatic carcinomas, 89% showed stenosis of the main pancreatic duct with proximal duct dilatation (Fig. 4-25). Concomitant bile duct obstructions are frequently seen (double duct sign) [20]. However, carcinomas limited to the tail of the pancreas, to the duct of Santorini, or to the uncinate process branches were difficult to diagnose with MRCP prospectively due to decreased spatial resolution (Figs. 4-26, 27,28).

MRCP has various advantages compared with ERCP (see section 1.2), and in our institution MRCP has replaced ERCP in the early diagnosis of pancreatic ductal adenocarcinomas (Figs. 4-29, 30).

Fig. 4-25 Pancreatic carcinoma

a. Single-slice MRCP shows stenosis in both Wirsung's duct and Santorini's duct (*arrow*) with proximal duct dilatation.

b. ERCP shows obstruction of the duct system, but proximal duct dilatation is not demonstrated.

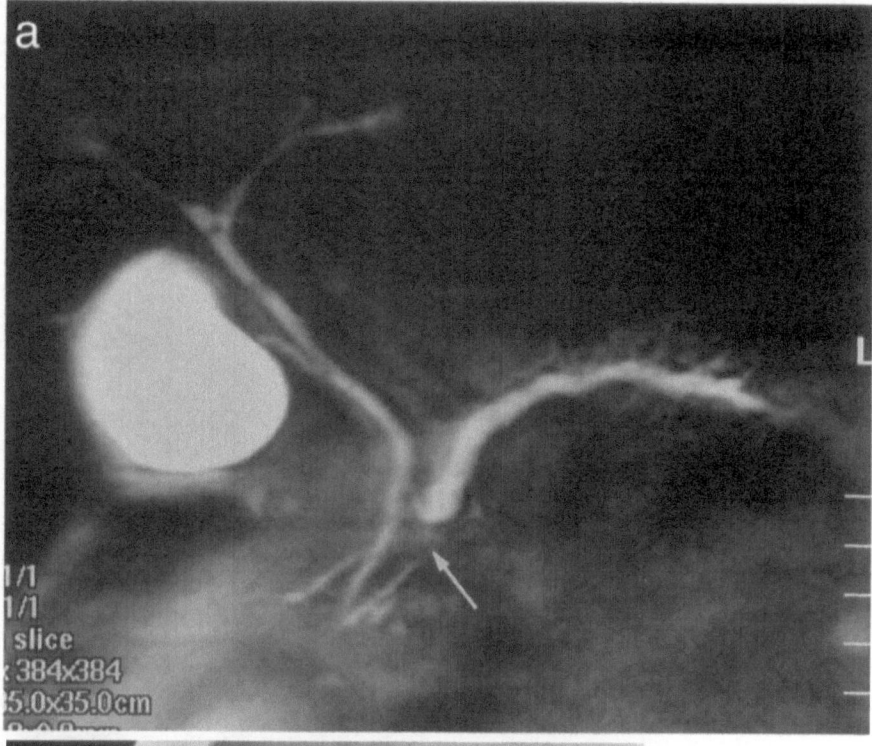

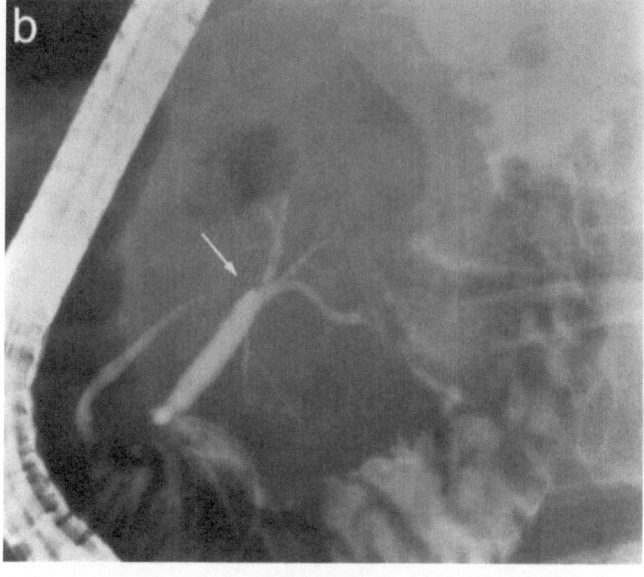

Fig. 4-26 Pancreatic carcinoma limited to the duct of Santorini

a. Three-dimensional MRCP with MIP shows nonvisualization of the duct of Santorini (*open arrow*) and stenosis of the distal common bile duct (*arrow*).

b. Histology revealed a pancreatic ductal adenocarcinoma that developed from the duct of Santorini with invasion of the distal common bile duct.

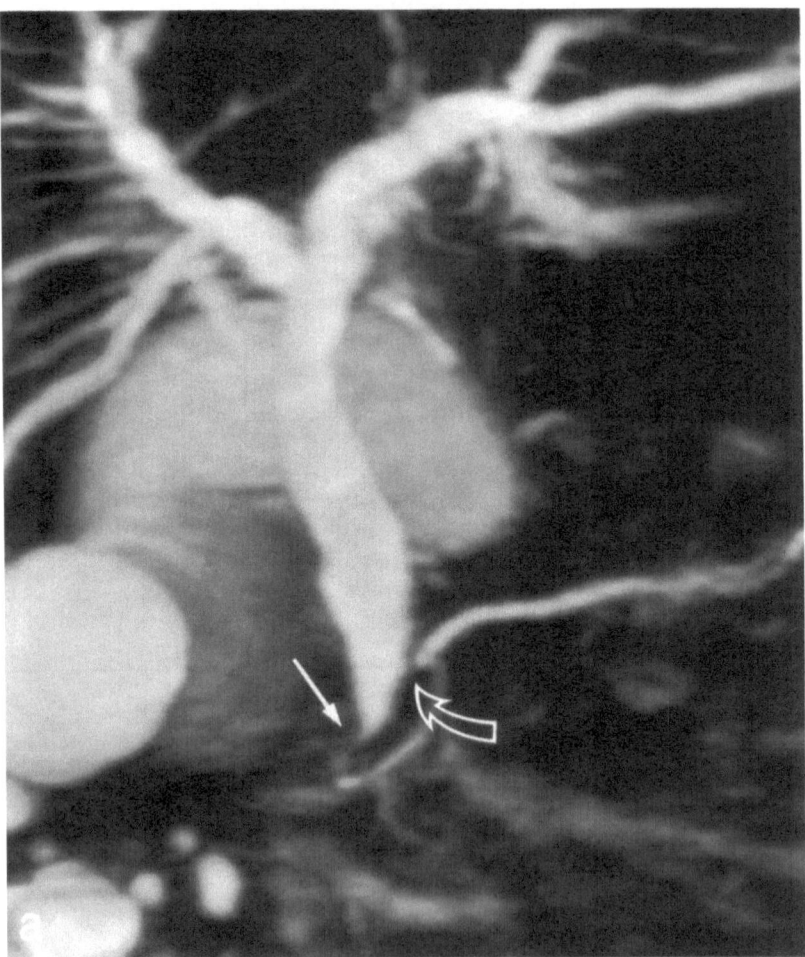

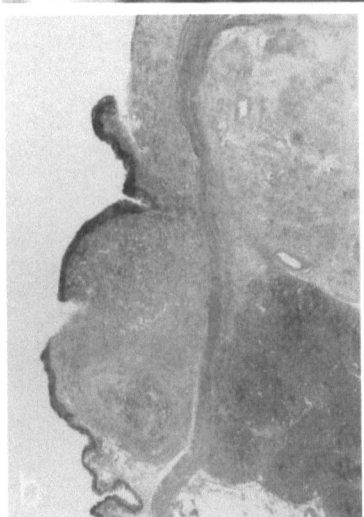

Fig. 4-27 Pancreatic carcinoma in the tail of the pancreas

a. Single-slice MRCP shows obstruction of the pancreatic duct in the tail (*arrow*).

b. ERCP also shows obstruction of the proximal pancreatic duct (*arrow*).

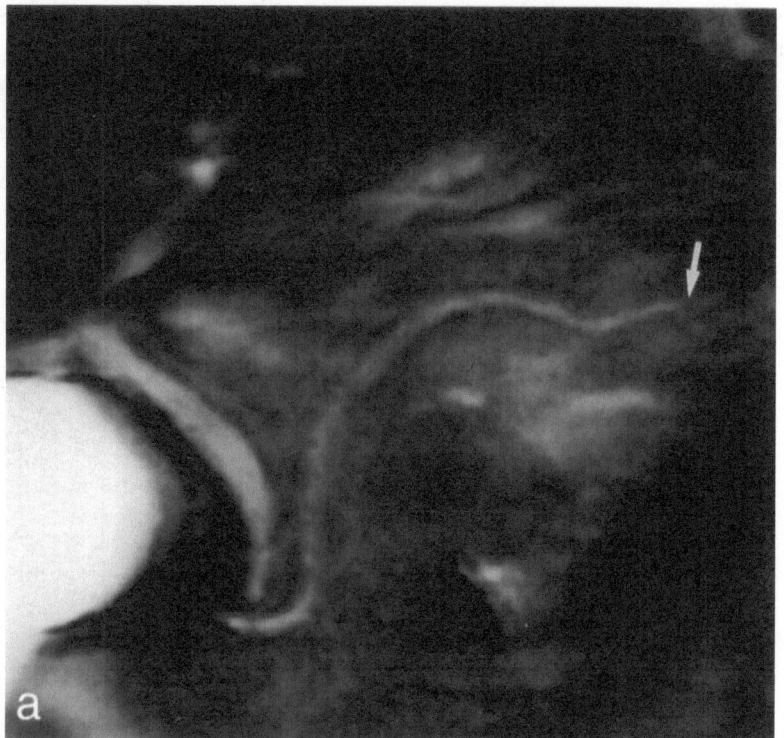

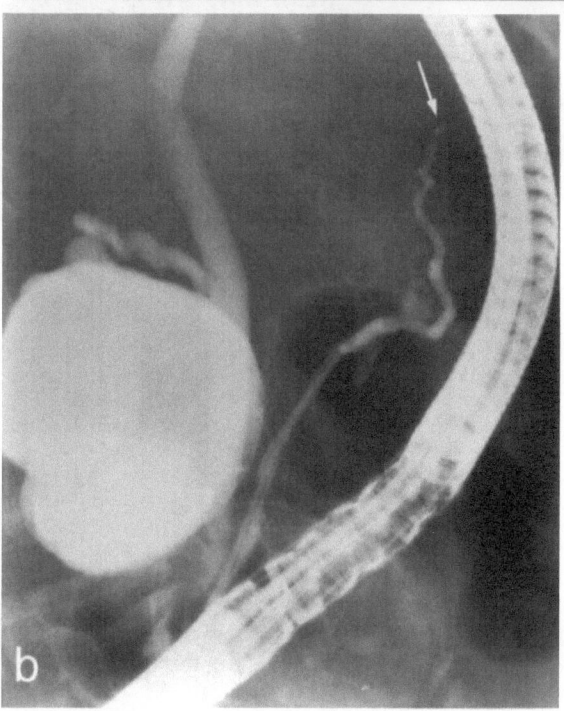

Fig. 4-28 Pancreatic carcinoma arising from the side branch

a. Single-slice MRCP shows stenosis of the distal common bile duct (*arrow*) and the normal pancreatic duct system.
b. ERCP shows stenosis of the distal common bile duct and no abnormality in the pancreatic duct system. Histology revealed pancreatic ductal adenocarcinoma arising from the small side branch infiltrating to the distal common bile duct.

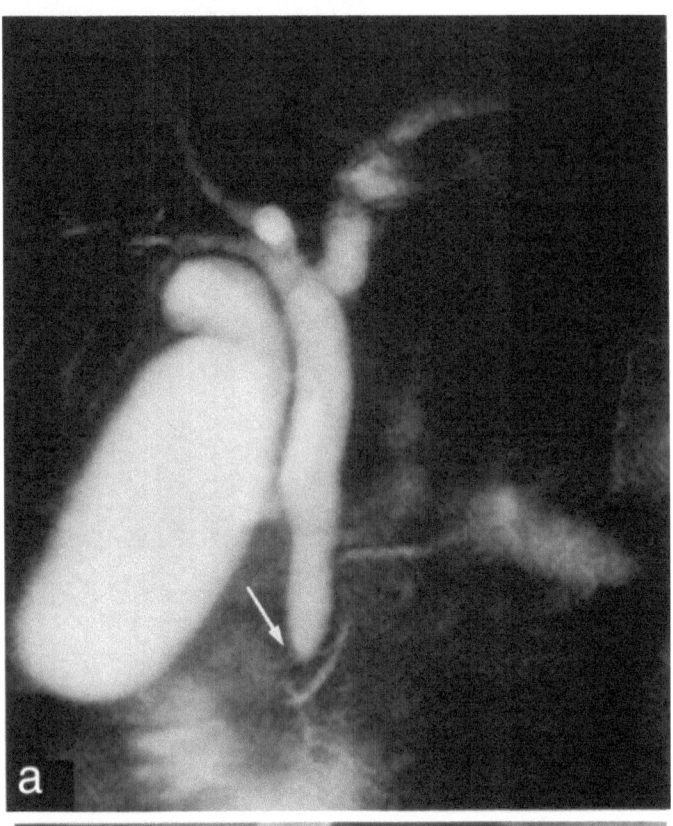

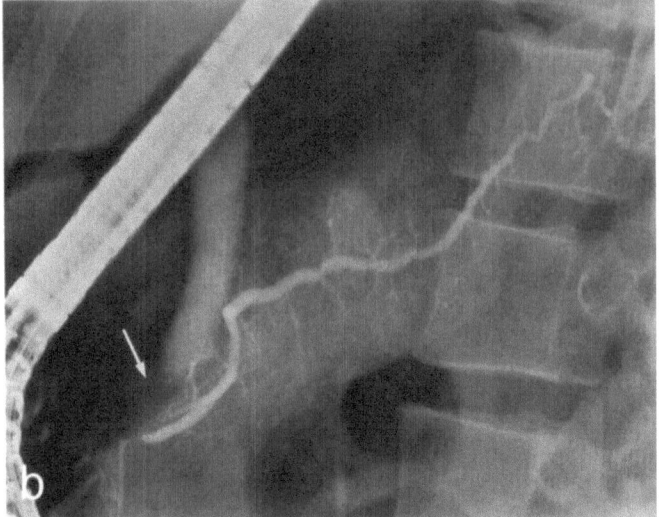

Fig. 4-29 Small pancreatic carcinoma 13 mm in diameter

a. Single-slice MRCP shows stenosis of the duct of Wirsung (*arrow*) with proximal duct dilatation.

b. ERCP shows the same findings as in MRCP.

c. Histology revealed a moderately differentiated tubular adenocarcinoma 13 mm in diameter.

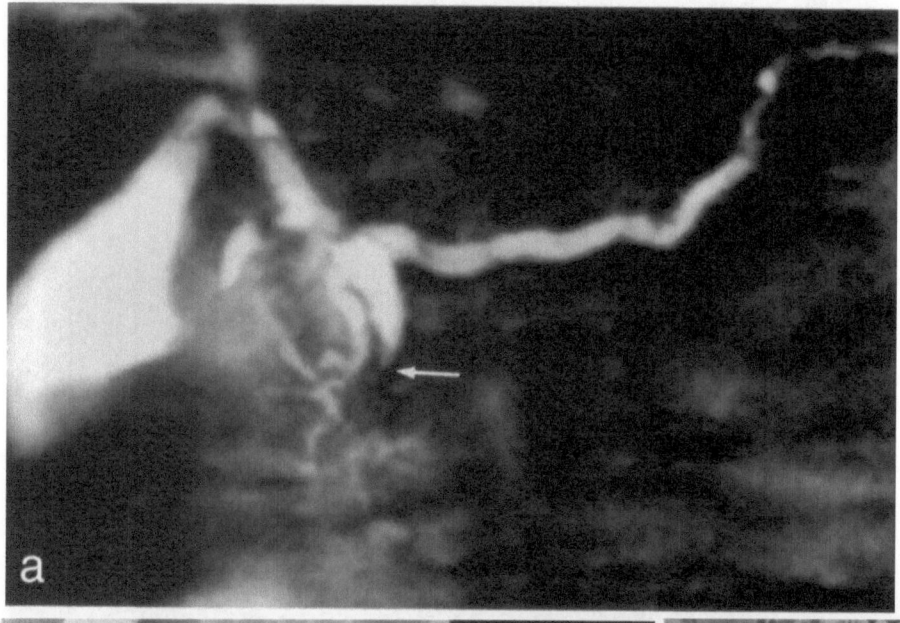

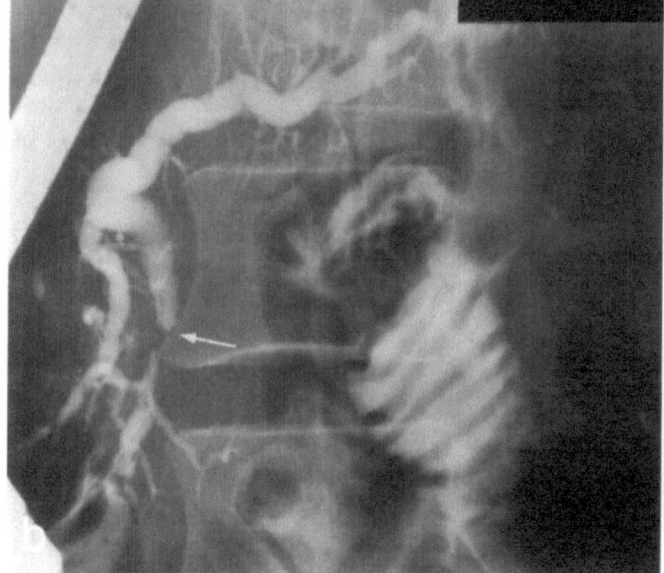

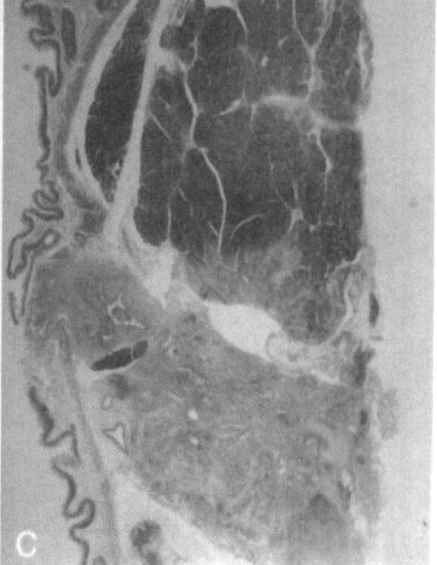

Fig. 4-30 Small pancreatic carcinoma 10 mm in diameter

a. Single-slice MRCP shows stenosis of the main pancreatic duct in the body (*open arrow*).

b. EUS (curved linear type) shows a small hypoechoic tumor (*arrowheads*) corresponding to duct stenosis.

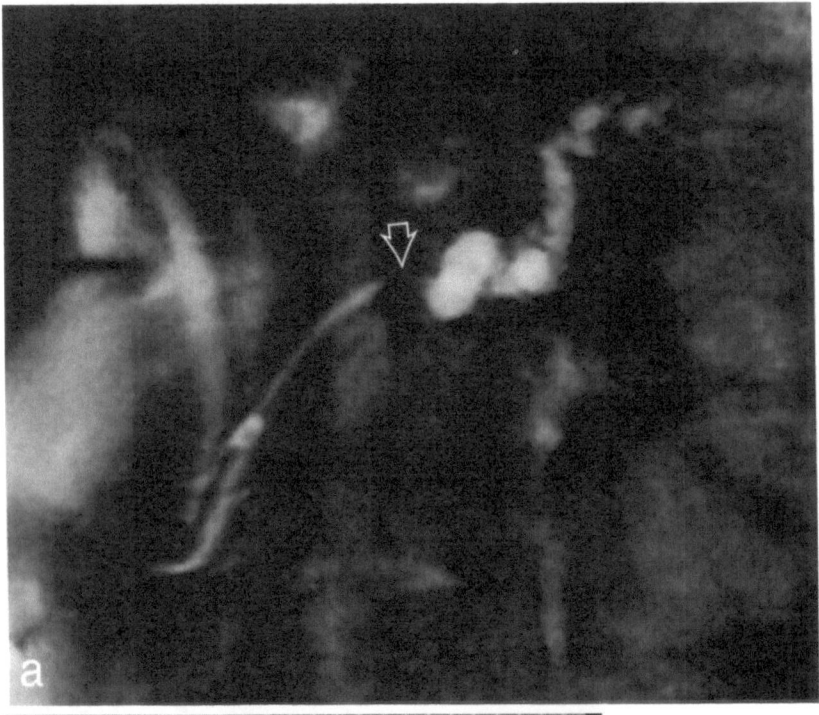

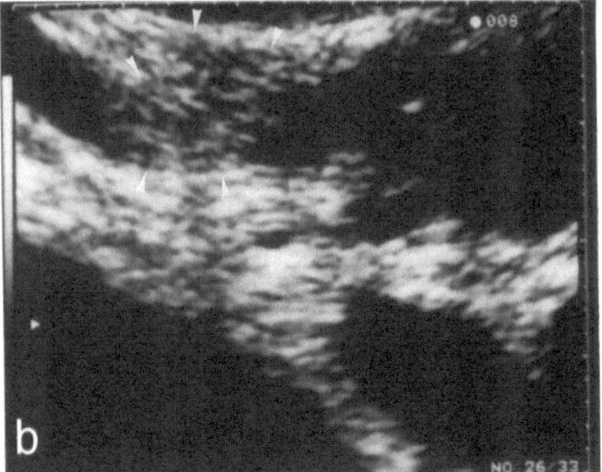

Fig. 4-30 *continued*

c. ERCP shows stenosis of the main pancreatic duct (*arrow*) and proximal duct dilatation.

d. Histology revealed a moderately differentiated tubular adenocarcinoma 10 mm in diameter.

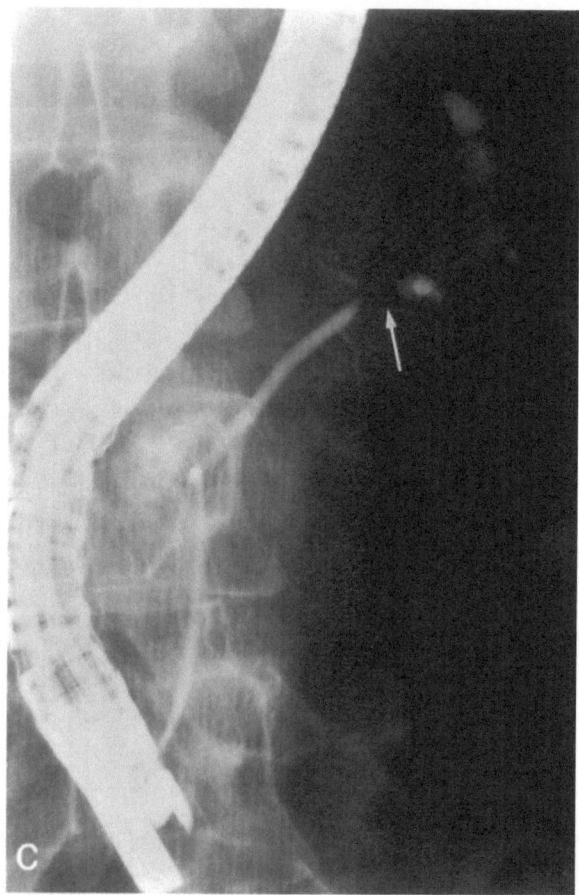

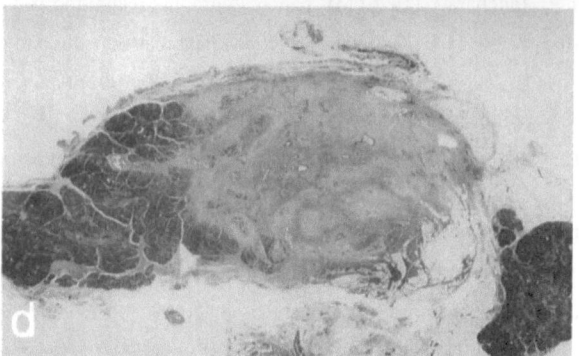

References

1. Miyazaki K, Yamashita Y, Tsuchigame T, et al (1996) MR cholangiopancreatography using HASTE (half-Fourier acquisition single-shot turbo spin-echo) sequences. AJR Am J Roentgenol 166: 1297-1303

2. Sai J, Ariyama J, Suyama M, et al (1999) MR cholangiopancreatography of the normal pancreatic duct system. Med Rev 69:27-31

3. Takehara Y, Ichijo K, Tooyama N, et al (1994) Breath-hold MR cholangiography with a long-echo-train fast spin-echo sequence and a surface coil in chronic pancreatitis. Radiology 192:73-78

4. Agha FD, Williams KD (1987) Pancreas divisum: incidence, detection, and clinical significance. Am J Gastroenterol 82:315-320

5. Delhaye M, Engelholm L, Cremer M (1985) Pancreas divisum: congenital anatomy variant or anomaly: contribution of endoscopic retrograde dorsal pancreatography. Gastroenterology 89:951-958

6. Bret PM, Reinhold C, Taourel P, et al (1996) Pancreas divisum: evaluation with MR cholangiopancreatography. Radiology 199:99-103

7. Barish M, Soto J, Ferrucci J (1997) Magnetic resonance pancreatography. Endoscopy 29:487-495

8. Sai J, Ariyama J, Suyama M, et al (1998) Clinical usage of MR cholangiopancreatography (in Japanese). Nippon Rinsho 56:2768-2772

9. Takehara Y (1996) MR pancreatography: technique and applications (1996) Top Magn Reson Imag 8:290-301

10. Wagner CW, Golladay ES (1988) Pancreas divisum and pancreatitis in children. Am Surg 54:22-26

11. Hirohashi S, Hirohashi R, Uchida H, et al (1997) Pancreatitis: evaluation with MR cholangiopancreatography in children. Radiology 203:411-415

12. Axon ATR, Classen M, Cotton PB, et al (1984) Pancreatography in chronic pancreatitis: international definitions. Gut 25:1107-1112

13. Sica GT, Braver J, Cooney MJ, et al (1999) Comparison of endoscopic retrograde cholangiopancreatography with MR cholangiopancreatography in patients with pancreatitis. Radiology 210:605-610

14. Soto JA, Barish MA, Yucel EK, et al (1995) Pancreatic duct: MR cholangio-pancreatography with a three-dimensional fast spin-echo technique. Radiology 196:459-464

15. Sai J, Ariyama J (1999) MRCP in the diagnosis of pancreatobiliary diseases: its progression and limitation (in Japanese). Jpn J Gastroenterol 96:259-265

16. Taylor AJ, Carmody TJ, Schmalz MJ, et al (1992) Filling defects in the pancreatic duct on endoscopic retrograde pancreatography. AJR Am J Roentgenol 159:1203-1208

17. Ferrucci JT, Wittenberg J, Black EB, et al (1979) Computed body tomography in chronic pancreatitis. Radiology 130:175-182

18. Neff CC, Simeone JF, Wittenberg J, et al (1984) Inflammatory pancreatic mass: problems in differentiating focal pancreatitis from carcinoma. Radiology 150:35-38

19. Yamaguchi K, Chijiwa K, Saiki S, et al (1996) Mass-forming pancreatitis masquerades as pancreatic carcinoma. Int J Pancreatol 20:27-35

20. Freeny PC, Bilbao MK, Katon RM (1976) Blind evaluation of endoscopic retrograde cholangiopancreatography (ERCP) in the diagnosis of pancreatic carcinoma: the double duct and other signs. Radiology 119:271-274

21. Ralls PW, Halls J, Renner I, et al (1980) Endoscopic retrograde cholangiopancreatography (ERCP) and pancreatic disease. Radiology 134:347-352

22. Ishihara T, Yamaguchi T, Tsuyuguchi T, et al (1996) Characteristic findings of endoscopic retrograde pancreatography in the diagnosis of chronic pancreatitis with inflammatory mass (in Japanese). Jpn J Gastroenterol 93:725-731

23. Yamaguchi K (1992) Pancreatic carcinoma associated with chronic pancreatitis. Int J Pancreatol 12:297-303

24. Cotton PB (1977) ERCP: progress report. Gut 18:316-341

25. Sai J, Ariyama J, Suyama M, et al (1999) MR cholangiopancreatography in the diagnosis of mass forming pancreatitis (in Japanese). J Bil Pancr 20:293-295

26. Takehara Y (1998) Can MRCP replace ERCP ? J Magn Reson Imaging 8:517-534.

27. Baron PL, Aabakken LE, Cole DJ, et al (1997) Differentiation of benign from malignant pancreatic masses by endoscopic ultrasound. Ann Surg Oncol 4: 639-643

28. Koito K, Namieno T, Nagakawa T, et al (1997) Inflammatory pancreatic masses: differentiation from ductal carcinomas with contrast-enhanced sonography using carbon dioxide microbubbles. AJR Am J Roentgenol 169:1263-1267

29. Stollfuss JC, Glatting G, Friess H, et al (1995) 2-(fluorine-18)-fluoro-2-deoxy-D-glucose PET in detection of pancreatic cancer: value of quantitative image interpretation. Radiology 195:339-344

30. Ikeda S, Matsumoto S, Yoshimoto H, et al (1984) Endoscopic balloon catheter pancreatography using catheter retaining technique. Usefulness of supine spot films by compression for complete evaluation of pancreatic ducts and branches (in Japanese). Stom Intes 19:1231-1242

31. Murakami K, Shiota K, Kubota T, et al (1994) A case of localized chronic pancreatitis masquerading as pancreatic cancer. J Jpn Pancr Soc 9:337-341

32. Becker V (1980) Sonderformen der chronischen Pankreatitis. Dtsch Arzteblatt 46:2711-2718

33. Stolte M, Weiss W, Volkholz H, et al (1982) A special form of segmental pancreatitis. Groove pancreatitis. Hepatogastroenterology 29:198-208

34. Yamaguchi K, Tanaka M (1992) Groove pancreatitis masquerading as pancreatic carcinoma. Am J Surg 163:312-316

35. Irie H, Honda H, Kuroiwa T, et al (1998) MRI of groove pancreatitis. J Comput Assist Tomogr 22:651-655

36. Sai J, Ariyama J, Suyama M, et al (1997) 3D-MR cholangiopancreatography in the diagnosis of cystic tumor of the pancreas (in Japanese). Endoscopia Digestiva 9:39-44

37. Sugiyama M, Atomi Y, Hachiya J (1998) Intraductal papillary tumors of the pancreas: evaluation with magnetic resonance cholangiopancreatography. Am J Gastroenterol 93: 156-159

38. Koito K, Namieno T, Ichimura T, et al (1998) Mucin-producing pancreatic tumors: comparison of MR cholangiopancreatography with endoscopic retrograde cholangiopancreatography. Radiology 208:231-237

39. Onaya H, Itai Y, Niitsu M, et al (1998) Ductectatic mucinous cystic neoplasms of the pancreas: evaluation with MR cholangiopancreatography. AJR Am J Roentgenol 171:171-177

40. Yeo CJ, Bastidas JA, Lynch-Nyhan A, et al (1990) The natural history of pancreatic pseudocysts documented by computed tomography. Surg Gyneco Obstet 170:411-417

41. van Sonnenberg E, Wittich GR, Casola G, et al (1989) Percutaneous drainage of infected and noninfected pancreatic pseudocyst: experience in 101 cases. Radiology 170:757-761

42. Bret PM, Reinhold C (1997) Magnetic resonance cholangiopancreatography. Endoscopy 29:472-486

43. Fulcher AS, Capps GW, Turner MA (1999) Thoracopancreatic fistula: clinical and imaging findings. J Comput Assist Tomogr 23:181-187

44. Warshaw AL, Compton CC, Lewandrowski K, et al (1990) Cystic tumors of the pancreas: New clinical , radiologic, and pathologic observations in 67 patients. Ann Surg 212:432-445

45. Prinson CW, Munson JL, Deveney CW (1990) Endoscopic retrograde cholangiopancreatography in the preoperative diagnosis of pancreatic neoplasms associated with cysts. Am J Surg 159:510-513

46. Martin I, Hannond P, Scott J, et al (1998) Cystic tumours of the pancreas. Br J Surg 85:1484-1486

47. Hulke J (1892) Pancreatic cyst. Lancet 2:1273

48. Jordan GL Jr (1960) Pancreatic cysts. In: Howard JM, Jordan GL Jr (eds) Surgical disease of the pancreas. Lippincott, Philadelphia, pp 283-320

49. Sperti C, Pasquali C, Costantino V, et al (1995) Solitary true cyst of the pancreas in adults. Int J Pancreatol 18:161-167

50. Ohhashi K, Murakami Y, Maruyama M, et al (1982) Four cases of mucous-secreting pancreatic cancer (in Japanese). Prog Dig Endosc 20:348-351

51. Yamada M, Kozuka S, Yamao K, et al (1991) Mucin-producing tumor of the pancreas. Cancer 68:159-168

52. Itai Y, Ohhashi K, Nagai H, et al (1986) Ductectatic mucinous cystadenoma and cystadenocarcinoma of the pancreas. Radiology 161:697-700

53. Yamaguchi K, Ogawa Y, Chijiwa K, et al (1996) Mucin-hypersecreting tumors of the pancreas: assessing the grade of malignancy preoperatively. Am J Surg 171:427-431

54. Sai J, Ariyama J (1992) Diagnosis and therapy of the cystic tumor of the pancreas (in Japanese). Jpn J Gastroenterol 89:1694

55. Usuki N, Okabe Y, Miyamoto T (1998) Intraductal mucin-producing tumor of the pancreas: diagnosis by MR cholangiopancreatography. J Comput Assist Tomogr 22: 875-879

56. Compagno J, Oertel JE (1978) Mucinous cystic neoplasms of the pancreas with overt and latent malignancy (cystadenocarcinoma and cystadenoma). A clinicopathological study of 41 cases. Am J Clin Pathol 69:573-580

57. Friedman AC, Lichtenstein JE, Dachman AH (1983) Cystic neoplasms of the pancreas. Radiological-pathological correlation. Radiology 149:45-50

58. Bergmann LS, Russell JC, Gladstone A, et al (1992) Cystadenomas of the pancreas. Am Surg 58:65-71

59. Prinson CW, Munson JL, Deveney CW (1990) Endoscopic retrograde cholangiopancreatography in the preoperative diagnosis of pancreatic neoplasms associated with cysts. Am J Surg 159:510-513

60. Minami M, Itai Y, Ohtomo K, et al (1989) Cystic neoplasms of the pancreas: comparison of MR imaging with CT. Radiology 171:53-56

61. Compagno J, Oertel JE (1978) Microcytic adenomas of the pancreas (glycogen-rich cystadenomas): a clinicopathologic study of 34 cases. Am J Clin Pathol 69:289-298

62. Johnson CD, Stephens DH, Charboneau JW, et al (1988) Cystic pancreatic tumors: CT and sonographic assessment. AJR Am J Roentgenol 151:1133-1138

63. Mathieu D, Guigui B, Valette PJ, et al (1989) Pancreatic cystic neoplasms. Radiol Clin N Am 27: 163-176

64. George DH, Murphy F, Michalski R, et al (1989) Serous cystadenoma of the pancreas: a new entity? Am J Surg Pathol 24:61-66

65. Kamei K, Funabiki T, Ochiai M, et al (1991) Multifocal pancreatic serous cystadenoma with atypical cells and focal perineural invasion. Int J Pancreatol 10:161-172

66. Gazelle GS, Mueller PR, Raafat N, et al (1993) Cystic neoplasms of the pancreas: evaluation with endoscopic retrograde pancreatography. Radiology 188:633-636

67. Freeney PC, Ball TJ (1978) Evaluation of endoscopic retrograde cholangiopancreatography and angiography in the diagnosis of pancreatic carcinoma. AJR Am J Roentgenol 130:683-691

68. Warshaw AL, Fernandez-del Castillo C (1992) Pancreatic carcinoma. N Engl J Med 326:455-465

69. Manabe T, Miyashita T, Ohshio G, et al (1988) Small carcinoma of the pancreas. Cancer 62:135-141

70. Nakaizumi A, Tatsuta M, Uehara H, et al (1992) A prospective trial of early detection of pancreatic cancer by ultrasonographic examination combined with measurement of serum elastase 1. Cancer 69:936-940

71. Ariyama J, Suyama M, Satoh K, et al (1998) Endoscopic ultrasound and intraductal ultrasound in the diagnosis of small pancreatic tumors. Abdom Imaging 23:380-386

72. Gabata T, Matsui O, Kadoya M, et al (1994) Small pancreatic adenocarcinomas: efficacy of MR imaging with fat suppression and gadolinium enhancement. Radiology 193:683-688

73. Tsuchiya R, Noda T, Harada N, et al (1985) Collective review of small carcinomas of the pancreas. Ann Surg 203:77-81

74. Ariyama J, Suyama M, Ogawa K, et al (1990) The detection and prognosis of small pancreatic carcinoma. Int J Pancreatol 7:37-47

75. Sai J, Ariyama J, Suyama M, et al (1998) MR cholangiopancreatography in the diagnosis of pancreatic ductal adenocarcinoma (in Japanese). J Bil Tract Pancr 19:39-44

76. Sai J, Ariyama J, Suyama M, et al (1995) Diagnosis of the small pancreatic cancer (in Japanese). J Bil Tract Pancr 16:219-225

77. Ariyama J, Shirakabe H, Ikenobe H, et al (1977) The diagnosis of the small resectable pancreatic carcinoma. Clin Radiol 28:437-444

Chapter 5　The Liver

5.1 Simple Hepatic Cysts

Simple hepatic cysts occur in 2.5% of the population, and the incidence increases with age [1]. They are seen more frequently in women, in the right lobe of the liver [1]. The cysts are usually unilocular, occasionally multiple, and are lined with a single layer of epithelium [2].

MRCP shows round, unilocular cysts with high signal intensity of cystic fluid.

Fig. 5-1 Multiple liver cysts

Single-slice MRCP shows multiple cysts in the liver.

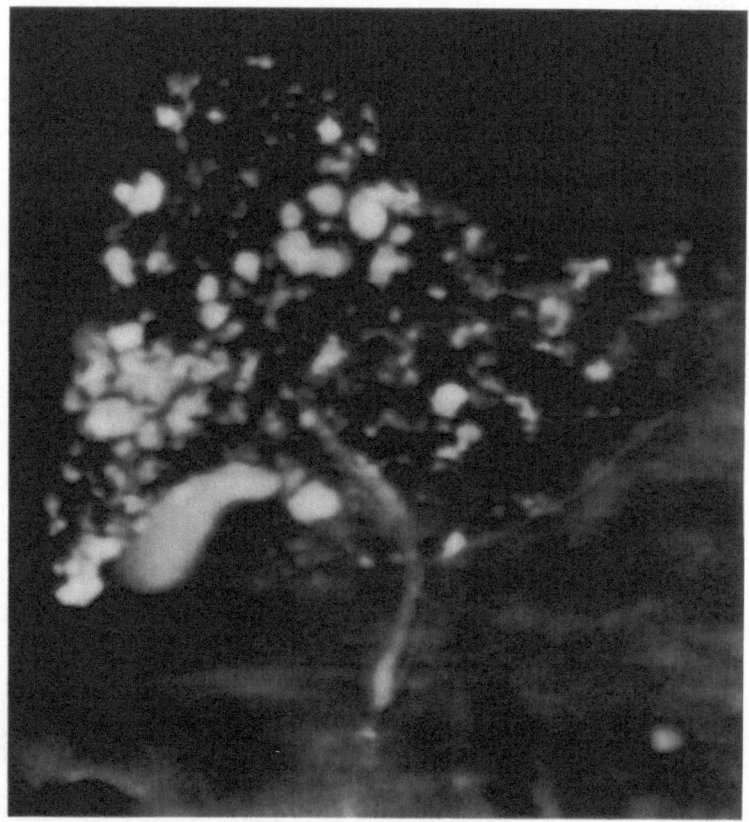

5.2 Hepatic Peribiliary Cysts and Polycystic Kidney

Hepatic peribiliary cysts are multiple small cysts, 0.2–2 cm in diameter. They are seen exclusively in the large portal tract, hepatic hilum, or both. Histologically, dilatations of the extramural peribiliary gland are seen. The cystic wall is composed of a single layer of columnar or cuboidal epithelium and a dense fibrous band without inflammation [3-5]. Hepatic peribiliary cysts are reported to be associated with idiopathic portal hypertension (100%), extrahepatic portal obstruction (100%), autosomal dominant polycystic kidney disease (ADPKD) (100%), systemic infection (68%), liver cirrhosis (50%), malignant metastases with obstructive jaundice (38%), and ascending cholangitis (31%) [6,7]. Von Meyenburg complexes and multiple cysts including intrahepatic cysts and peribiliary cysts may exist together, especially in patients with ADPKD [8].

MRCP demonstrates small cysts along either side or both sides of the portal veins and the bile ducts in the hepatic hilum to the third-level branch. MRCP clearly demonstrates the distribution of hepatic peribiliary cysts and the relationship with the bile duct. Differentiation from Caroli's disease is feasible, the latter showing multiple cystic lesions in connection with the intrahepatic biliary tree throughout the liver [9,10].

Fig. 5-2 Hepatic peribiliary cysts and polycystic kidney

Single-slice MRCP shows multiple small cysts along the portal vein and the bile ducts in the hepatic hilum, associated with polycystic kidney.

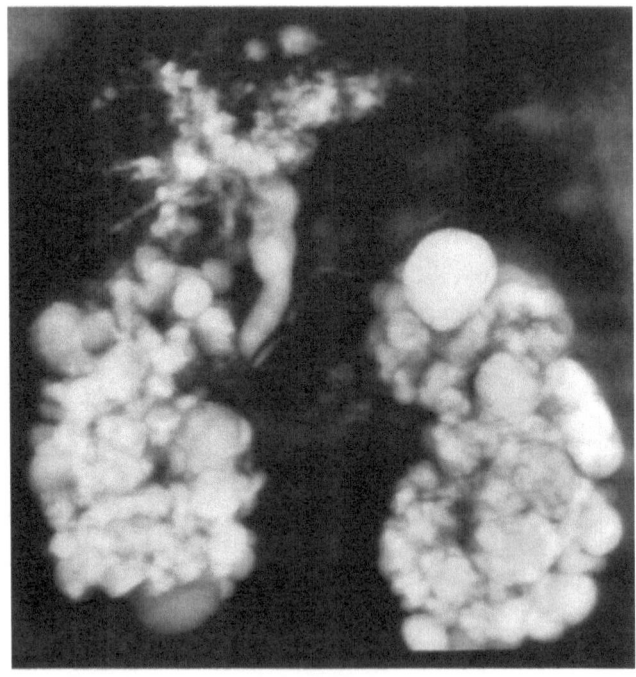

5.3 Von Meyenburg Complexes (Microhamartomas)

Von Meyenburg complex (VMC) is usually asymptomatic, diagnosed incidentally, and associated with autosomal dominant polycystic kidney diseases (ADPKDs) in 11% of cases. Conversely, VMCs are found in 97% of adults with ADPKD [8].

VMC consists of a variable number of irregular bile ductal structures [8,11]. The lumina of these bile ductules often contain bile and may communicate with the normal terminal bile ducts in the neighboring portal tract area as part of the still-functioning bile ducts [12].

MRCP shows multiple or numerous intrahepatic tiny (usually less than 5 mm) cystoid lesions with high signal intensity, reflecting the fluid nature of VMC. The lesions are scattered throughout the liver up to the subcapsular areas, and are irregular in shape [13,14]. They are located near the portal tract, but compared with hepatic peribiliary cysts, they are more diffuse and do not show a close relationship with the larger portal tract [5] (see section 5.2).

Fig. 5-3 Von Meyenburg complex

a. US shows multiple comet-tail echoes and tiny cysts throughout the liver.
b. Single-slice MRCP shows numerous tiny, irregular high-signal intensity spots, scattered throughout the liver.

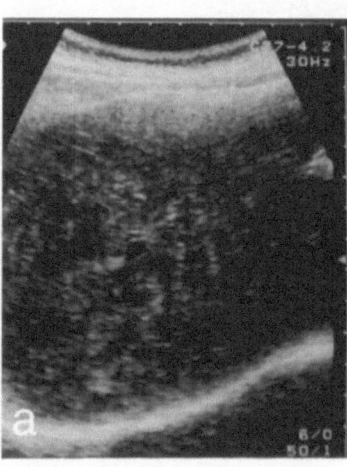
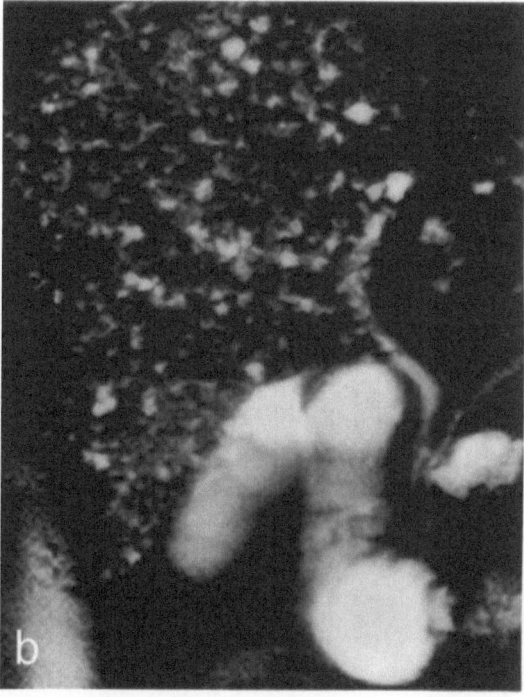

5.4 Biliary Cystadenoma and Cystadenocarcinoma

Biliary cystadenoma and cystadenocarcinoma are considered as the same entity, and transformation of cystadenoma to cystadenocarcinoma is recognized. This rare tumor occurs predominantly in middle-aged women. The tumors are usually multilocular, containing mucus, and lined by mucin-producing epithelium. Papillary protrusions are frequently present in the tumors. No communication can be seen between the tumors and the large biliary tract [15-17].

Cross-sectional images, such as US, CT, and axial MRI, are essential to diagnose the morphology and vascularity of the tumor[18]. However, MRCP demonstrates multilocular cystic tumor with high signal intensity and intrahepatic bile duct involvement or compression associated with tumor. By reviewing the source images of MIP, papillary protrusions within the cyst can be depicted.

Differentiation between cystadenoma and cystadenocarcinoma is often difficult by diagnostic imaging.

Fig. 5-4 Biliary cystadenocarcinoma

a. Single-slice MRCP shows a multilocular cyst (*bold arrow*). Filling defects (*open arrow*) are observed in the cyst.

b. Source image of 3D-MRCP depicts papillary protrusions inside the cyst (*arrowheads*).

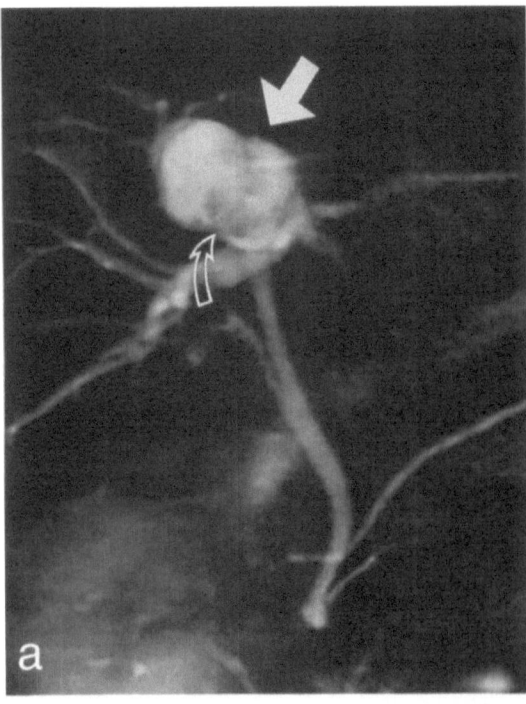
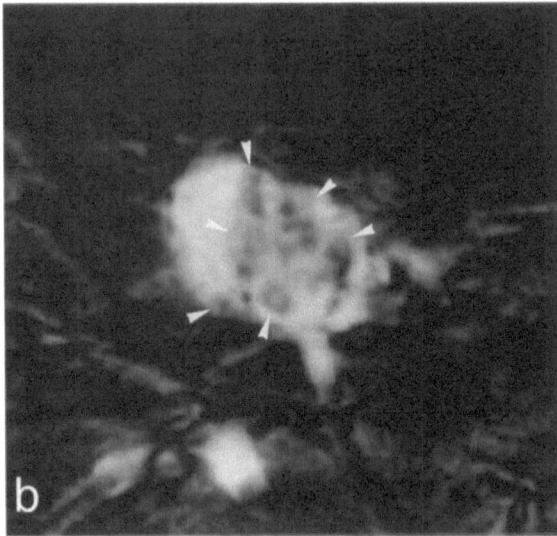

5.5 Intrahepatic Cholangiocarcinoma

Intrahepatic cholangiocarcinoma originates in the small or peripheral intrahepatic bile ducts. Hilar bile duct carcinomas (Klaskin's tumors) are excluded. The relative frequency of intrahepatic cholangiocarcinoma among all primary liver cancer is 25% in the United States [19] and 5.4% in Japan [20]. The tumors are firm, white masses and contain a large amount of fibrous tissue. They are adenocarcinomas with glandular appearance, resemble bile duct epithelium, and sometimes have a papillary arrangement. They are usually desmoplastic, and mucin formation is frequently observed [21-24].

Cross-sectional images, such as US, CT, and MRI, are essential in the diagnosis of intrahepatic cholangiocarcinomas [25,26]. However, the tumors involve the intrahepatic or proximal extrahepatic bile duct in 64% [27] of cases; focal stenotic lesions are most common, and intraluminal polypoid lesions or diffuse sclerosing lesions are less common [27]. Therefore, MRCP may aid diagnosis by depicting the stenosis, dilatation, filling defects, and displacement of the intrahepatic bile duct associated with tumors.

Fig. 5-5 Intrahepatic cholangiocarcinoma

a. Contrast-enhanced CT shows a low-density mass with peripheral enhancement in the left lobe of the liver. Atrophy and proximal bile duct dilatation are also observed in the left lobe.

b. Three-dimensional MRCP with MIP shows stenosis (*arrow*) and proximal duct dilatation of the left hepatic duct associated with the tumor.

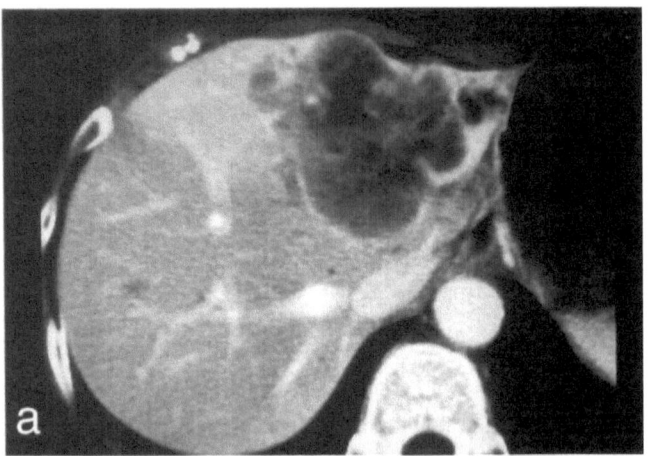

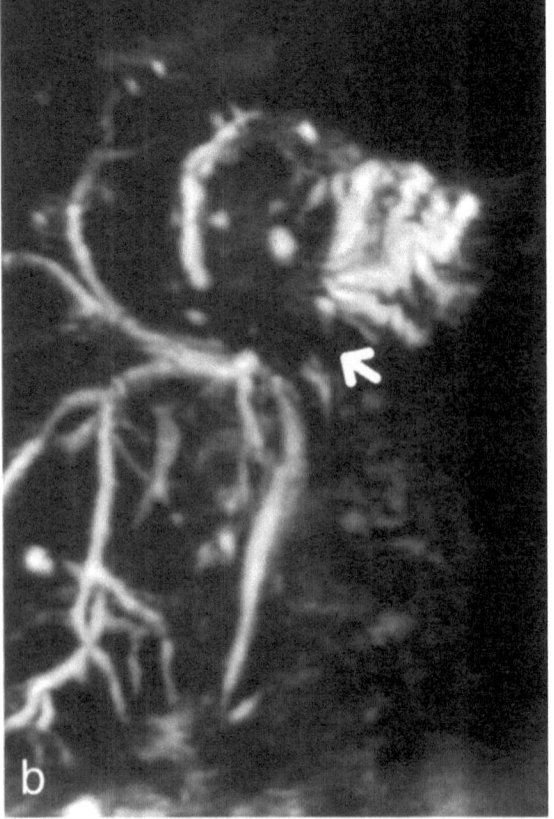

Fig. 5-6 Small intrahepatic cholangiocarcinoma

a. CT shows slight dilatation of the intrahepatic bile ducts in the left lobe (*arrow*).

b. Single-slice MRCP shows stenosis and proximal dilatation of the intrahepatic bile duct in the left lobe (*arrow*).

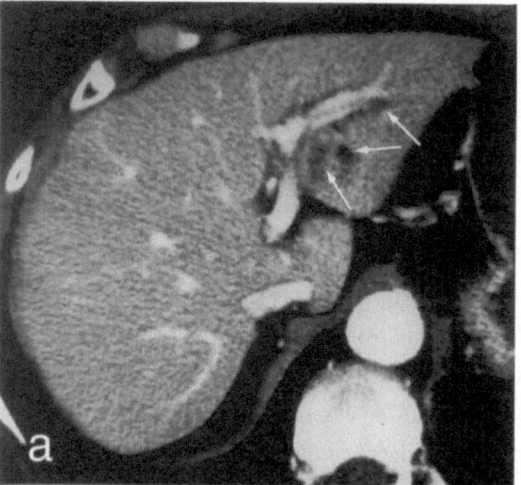

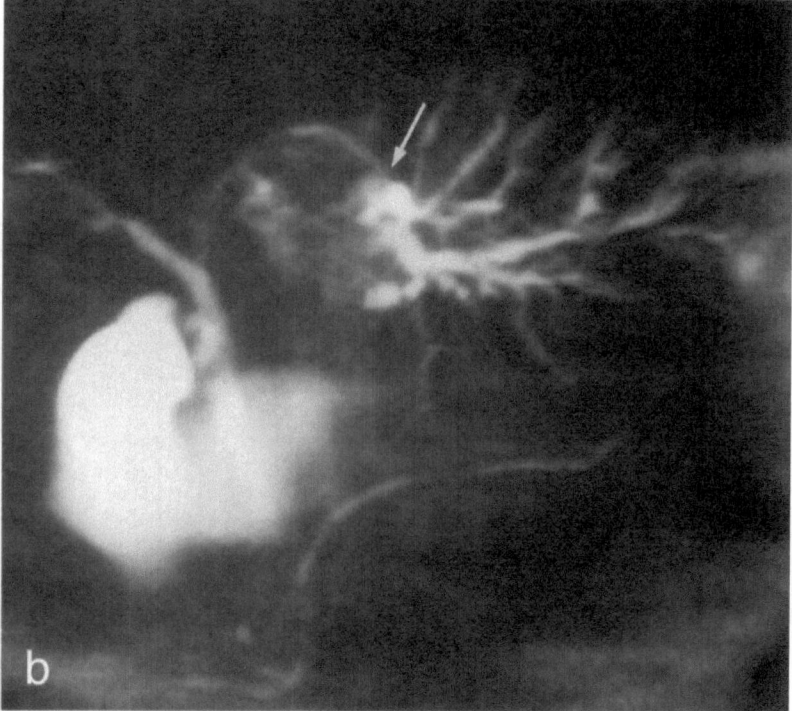

5.6 Hepatocellular Carcinoma

MRCP can provide additional information about the biliary tract in the diagnosis of hepatocellular carcinoma (HCC), especially before surgery, because bile duct involvement is reported in 12% of HCC cases [28,29]. MRCP demonstrates intraluminal filling defects within the major bile ducts, prestenotic dilatation of the bile ducts, or displacement and stretching of the extrahepatic bile duct or intrahepatic bile duct by the tumor mass [28].

Fig. 5-7 Hepatocellular carcinoma

a. CT shows a large HCC in the right lobe of the liver.
b. Single-slice MRCP shows displacement and stretching of the intrahepatic bile duct (*arrow*) by the tumor (*bold arrow*).

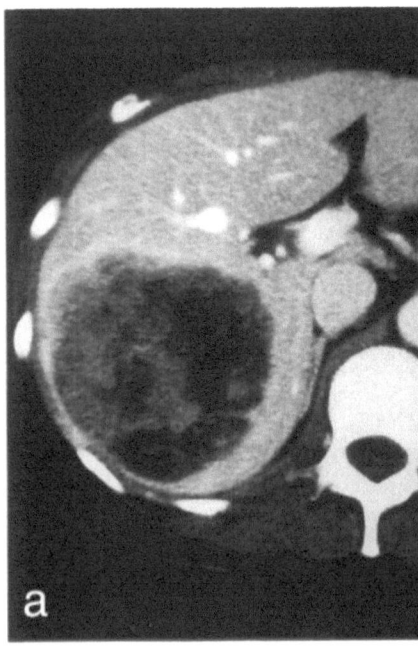
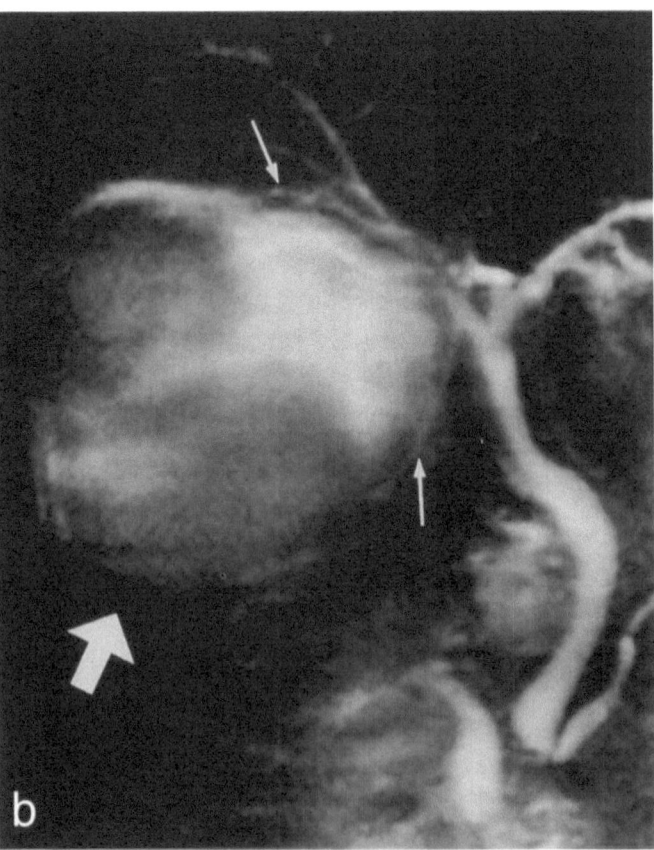

5.7 Pyogenic Liver Abscesses

In pyogenic liver abscesses, bacterial infection may occur via the biliary tract in 33%, the portal vein in 30%, the hepatic artery in 15%, and by direct extension in 15%. The abscesses may also occur as a result of trauma or may be idiopathic. Infection via the bile duct is the most common cause [30]; however, an increase in idiopathic abscesses has been observed in recent years.

Abscesses associated with biliary tract diseases are multiple and involve both hepatic lobes in approximately 90% of cases. Abscesses from the portal venous system are usually solitary, with 65% of abscesses occurring in the right lobe, 12% in the left lobe, and 23% in both lobes [31].

MRCP can depict the relationship between abscesses and the biliary tract, and provide additional information about the biliary tract, gallstones, pneumobilia, or bile duct stricture including malignancy.

Fig. 5-8 Pyogenic liver abscess

a. Single-slice MRCP shows a round mass with high signal intensity and an indefinite border in the medial segment of the liver. The internal structure is not homogeneous (*bold arrow*). Stones (*arrows*) are present in the common bile duct.
b. Percutaneous drainage shows liver abscess.

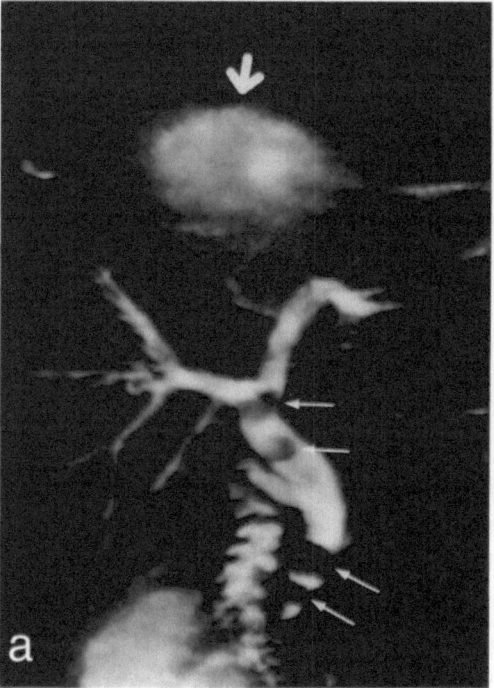
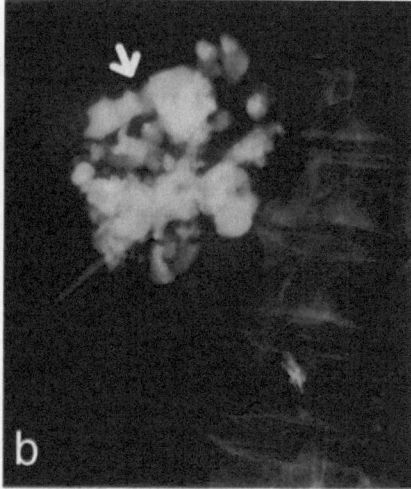

References

1. Gaines PA, Sampson MA (1989) The prevalence and characterization of simple hepatic cysts by ultrasound examination. Br J Radiol 62:335-337

2. Ishak KG (1987) New developments in diagnostic liver pathology. In: Farber E, Philips MJ, Kaufman N (eds) Progress of liver diseases. Williams & Wilkins, Baltimore, pp 223-373

3. Nakanuma Y, Kurumaya H, Ohta G (1984) Multiple cysts in the hepatic hilum and their pathogenesis: a suggestion of periductal gland origin. Virchows Arch A Pathol Anat 404:341-350

4. Terada T, Minato H, Nakanuma Y, et al (1992) Ultrasound visualization of hepatic peribiliary cysts: a comparison with morphology. Am J Gastroenterol 87:1499-1502

5. Itai Y, Ebihara R, Tohno E, et al (1994) Hepatic peribiliary cysts: multiple tiny cysts within the larger portal tract, hepatic hilum, or both. Radiology 191:107-110

6. Terada T, Nakanuma Y (1990) Pathological observations of intrahepatic peribiliary glands in 1,000 consecutive autopsy livers. III. Survey of necroinflammation and cystic dilatation. Hepatology 12:1229-1233

7. Itai Y, Ebihara R, Eguchi N, et al (1995) Hepatobiliary cysts in patients with autosomal dominant polycystic kidney disease: prevalence and CT findings. AJR Am J Roentgenol 164:339-342

8. Redston MS, Wanless IR (1996) The hepatic von Meyenburg complex with hepatic and renal cysts. Mod Pathol 9:233-237

9. Pavone P, Laghi A, Catalano C, et al (1996) Caroli's disease: evaluation with MR cholangiography. AJR Am J Roentgenol 166:216-217

10. Asselah T, Ernst O, Sergent G, et al (1998) Caroli's disease: a magnetic resonance cholangiopancreatography diagnosis. Am J Gastroenterol 93:109-110

11. Chung EB (1970) Multiple bile-duct hamartomas. Cancer 26:287-296

12. Ohta W, Ushio H (1984) Histological reconstruction of a von Meyenburg's complex on the liver surface. Endoscopy 16:71-74

13. Luo TY, Itai Y, Eguchi N, et al (1998) Von Meyenburg complex of the liver: imaging findings. J Comput Assist Tomogr 22:372-378

14. Eisenberg D, Hurwitz L, Yu AC (1986) CT and sonography of multiple bile-duct hamartomas simulating malignant liver disease. AJR Am J Roentgenol 147:279-280

15. Ishak KG, Willis GW, Cummins SD, et al (1977) Biliary cystadenoma and cystadenocarcinoma: report of 14 cases and review of the literature. Cancer 38:322-328

16. Nakajima T, Sugano I, Matsuzaki O, et al (1992) Biliary cystadenocarcinoma of the liver: a clinicopathological and histochemical evaluation of nine cases. Cancer 69:2426-2432

17. Wheeler DA, Edmondson HA (1985) Cystadenoma with mesenchymal stroma (CMS) in the liver and bile ducts. A clinicopathologic study of 17 cases, 4 with malignant change. Cancer 56:1434-1445

18. Korobkin MT, Stephens DH, Lee JKT, et al (1989) Biliary cystadenoma and cystadenocarcinoma: CT and sonographic findings. AJR Am J Roentgenol 153:507-511

19. Edmondson HA, Steiner PE (1954) Primary carcinoma of the liver. A study of 100 cases among 48,900 necropsies. Cancer 7:462-503

20. Okuda K and Liver Cancer Study Group of Japan (1980) Primary liver cancer. Cancer 45:2663-2669

21. Okuda K, Kudo Y, Okazaki N, et al (1977) Clinical aspects of intrahepatic bile duct carcinoma including hilar carcinoma. A study of 57 autopsy-proven cases. Cancer 39:232-246

22. Mori W, Nagasako K (1976) Cholangiocarcinoma and related lesions. In: Okuda K, Peters RL (eds) Hepatocellular carcinoma. Wiley, New York, pp 227-246

23. Sugihara S, Kojiro M (1987) Pathology of cholangiocarcinoma. In: Okuda K, Ishak KG (eds) Neoplasms of the liver, Springer, Tokyo, pp 143-158

24. Ros PR, Buck JL, Goodman ZD, et al (1988) Intrahepatic cholangiocarcinoma: radiologic-pathologic correlation. Radiology 167:689-693

25. Itai Y, Araki T, Furui S, et al (1983) Computed tomography of primary intrahepatic biliary malignancy. Radiology 147:485-490

26. Vilgrain V, Van Beers BE, Flejou JF, et al (1997) Intrahepatic cholangiocarcinoma: MRI and pathologic correlation in 14 patients. J Comput Assist Tomogr 21:59-65

27. Nichols DA, MacCarty RL, Gaffey TA (1983) Cholangiographic evaluation of bile duct carcinoma. AJR Am J Roentgenol 141:1291-1294

28. Lee NW, Wong KP, Siu KF, et al (1984) Cholangiography in hepatocellular carcinoma with obstructive jaundice. Clin Radiol 35:119-123

29. Kojiro M, Kawabata K, Kawano Y, et al (1982) Hepatocellular carcinoma presenting as intrabile duct tumor growth: a clinicopathological study of 24 cases. Cancer 15:2144-2147

30. Frey CF, Zhu Y, Suzuki M, et al (1989) Liver abscesses. Surg Clin N Am 69:259-271

31. Pitt HA (1991) Liver abscess. In: Zuidema GD (ed) Shackelford's surgery of the alimentary tract (3rd edn). Saunders, Philadelphia, pp 443-465

Chapter 6 Applications of Magnetic Resonance Cholangiopancreatography

6.1 Dynamic MRCP by Secretin Injection

Single thick slice MRCP using half-Fourier fast-spin echo techniques (HASTE, FASE, SSFSE) acquires a pancreatobiliary image within a few seconds and permits real-time observation of pancreatic juice secretion into the duodenum after intravenous injection of secretin [1-4]. A time-intensity curve can be plotted by calculating the total intensity of the gastric and duodenal fluid in various phases before and after secretin injection [4].

Fig. 6-1 Dynamic MRCP (two-dimensional)

a. MRCP before and after intravenous injection of secretin. The time-course of pancreatic juice secretion is depicted.
b. Comparison of the MRCP-derived time-intensity curve of pancreatic juice secretion and endoscopic sampling of pancreatic juice. A good correlation is obtained between the two methods.

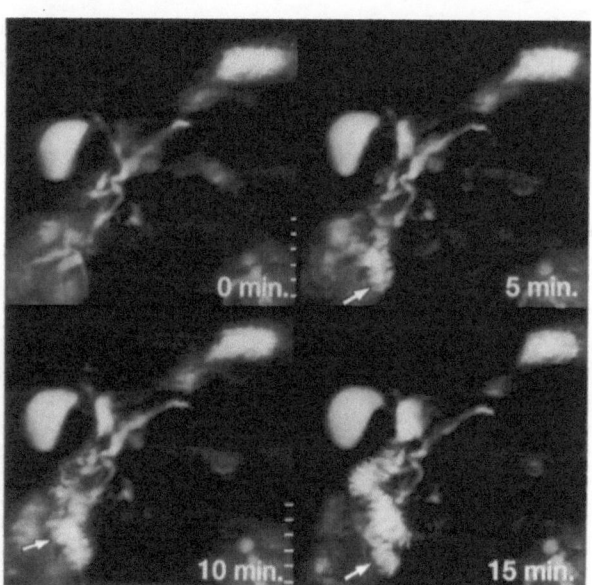

Time Course of the Pancreatic Juice Secretion

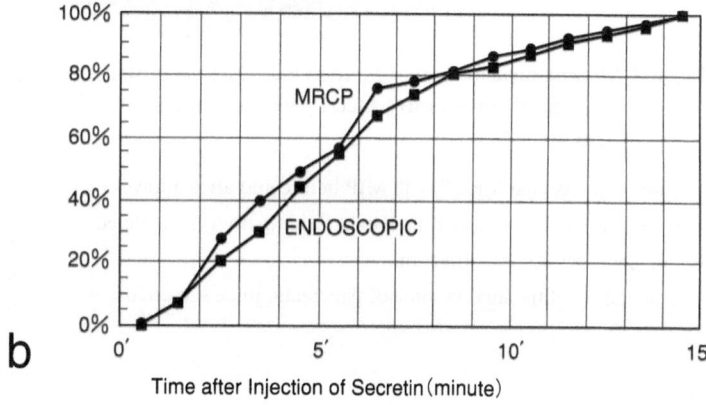

6.2 Measurement of Volume of Pancreatic Juice Secretion

The total volume of pancreatic juice secretion after intravenous secretin injection can be determined noninvasively by measuring the volumes of gastrointestinal fluid before and after injection of secretin and calculating the difference (Fig. 6-2). Three-dimensional images must be generated for calculating the volume of gastrointestinal fluid. Figure 6-3 indicates the methods of calculation. From the source images, the total signal intensity inside the gastrointestinal tract is calculated using the region-growing method (Fig. 6-4). Signal intensity then weights the volume by voxel summation, and the total volume of the fluid is obtained (Fig. 6-5). Source images should be reviewed at intervals of several millimeters to minimize partial volume effects.

We acquired 40–60 source images with a 384 × 384 matrix and 1 mm slice thickness using 3D-half-Fourier RARE sequence (3D-FASE sequence). Excellent results were achieved in phantom experiments and in clinical measurements (Tables 6-1, 6-2) [4].

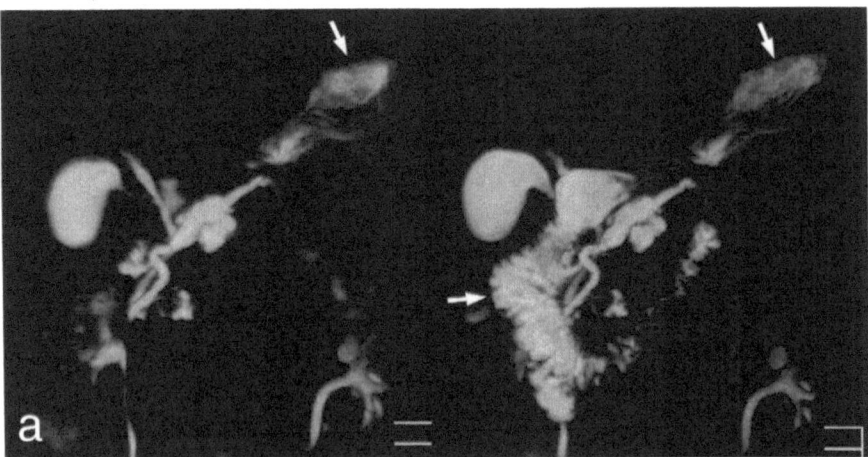

Fig. 6-2 Volume measurement of pancreatic juice secretion in a patient with intraductal papillary mucinous tumor

a. Three-dimensional MRCP with MIP before and after intravenous secretin injection. Gastrointestinal fluid (*arrows*) is depicted before and after secretin injection.
b. Calculation of the total volume of pancreatic juice secretion.

b

$$V_{total} = V_{15min.} - V_{0min.}$$

V_{total}: total volume of pancreatic juice secretion 15min. after secretin injection

$V_{0min.}$: total volume of gastrointestinal fluid before secretin injection

$V_{15min.}$: total volume of gastrointestinal fluid 15min. after secretin injection

Fig. 6-3 Volume measurement using source images of 3D-MRCP

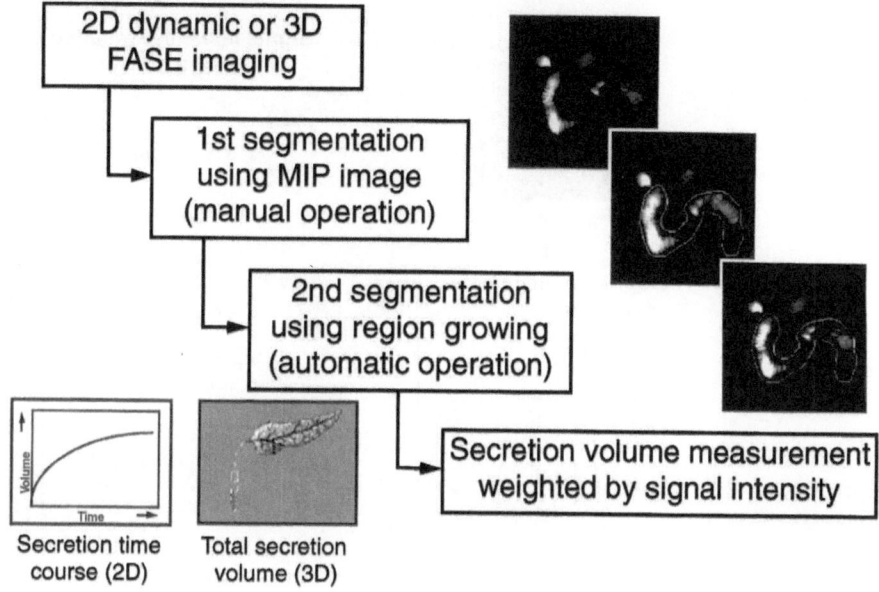

Fig. 6-4 Region-growing method

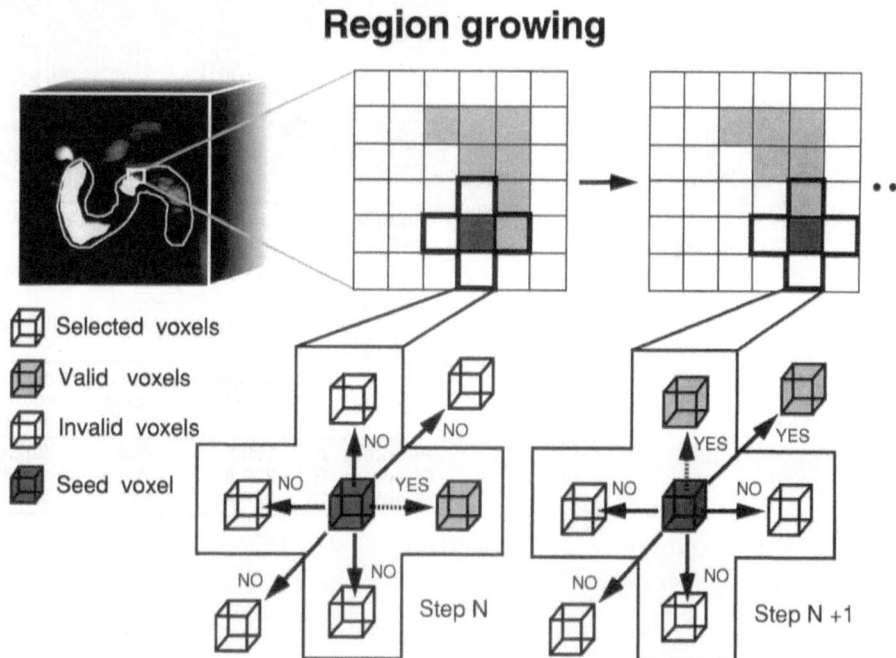

Fig. 6-5 Signal-intensity weighted volume calculation

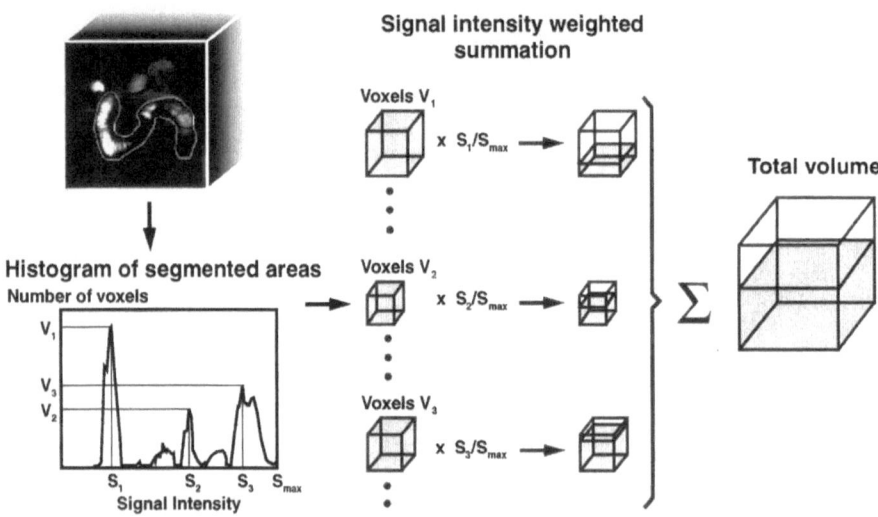

Signal intensity weighted volume calculation

Table 6-1 Phantom experiments of volume determination

The error in volume determination is increased as the phantom structure becomes more complicated. The maximum error range is within 10%. The major source of error is probably partial volume effects.

Volume measurement dependency on the phantom structure.

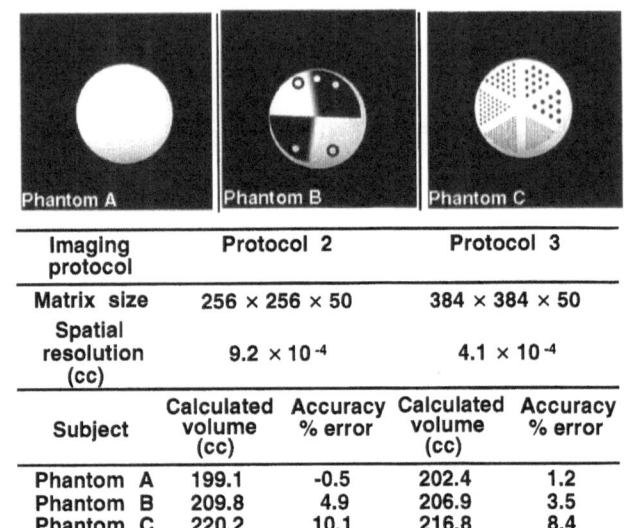

Imaging protocol	Protocol 2		Protocol 3	
Matrix size	$256 \times 256 \times 50$		$384 \times 384 \times 50$	
Spatial resolution (cc)	9.2×10^{-4}		4.1×10^{-4}	
Subject	Calculated volume (cc)	Accuracy % error	Calculated volume (cc)	Accuracy % error
Phantom A	199.1	-0.5	202.4	1.2
Phantom B	209.8	4.9	206.9	3.5
Phantom C	220.2	10.1	216.8	8.4

**Table 6-2 Comparison of MRCP measurement and endoscopic
pancreatic juice sampling**

Five patients with intraductal papillary mucinous tumors who required
endoscopic pancreatic juice cytology were studied. The mean difference between the MRCP measurement and endoscopic sampling was
5.8 ± 4.2 ml (mean ± SD).

	Endoscope Sampling	MRCP Measurement	Difference
Case 1	24.6	34.1	9.5
Case 2	45.0	36.2	8.8
Case 3	26.0	23.9	2.4
Case 4	8.0	16.2	8.2
Case 5	33.2	33.3	0.1
Mean			5.8
SD			4.2

6.3 MR Virtual Cholangiopancreatoscopy

MR virtual cholangiopancreatoscopy can be generated by setting a signal-intensity threshold for the image data of 3D-MRCP and reconstructing the images using the surface rendering method (Fig. 6-6). In this method, a viewpoint and a light source are specified in the lumen for displaying the image. Objects located near the viewpoint appear larger, while those located far from it appear smaller. The light source provides illumination and shading, producing virtual images of the pancreatic duct and biliary tract. To obtain high-resolution images, we acquired 40–60 source images with a 384×384 matrix and 1 mm slice thickness using 3D-half-Fourier RARE sequence (3D-FASE sequence).

This method is still in the early stage of development and has a number of limitations such as different threshold settings resulting in markedly different images, inadequate spatial resolution, no color display (e.g., of redness), and hole artifacts [5]. Technical refinement is required for further development of this technique.

Fig. 6-6 Gallbladder polyp

a. Three-dimensional image by surface rendering. The viewpoint is set inside the gallbladder, and the viewing direction is from the body toward the fundus (*arrows*).

b. MR virtual cholangioscopy shows a polyp inside the gallbladder (*arrow*).

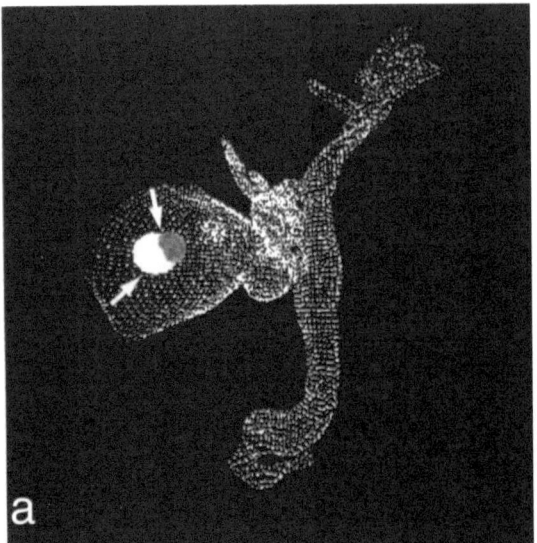
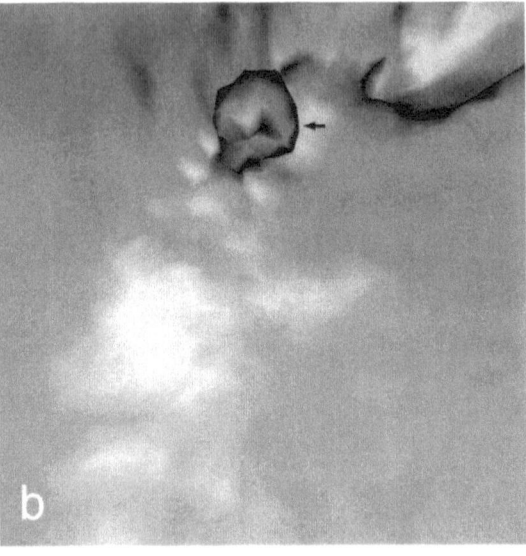

Fig. 6-7 Intraductal papillary mucinous tumor

a. Three-dimensional MRCP with MIP shows dilatation of the main pancreatic duct (*open arrow*) and cystic dilatation of a side branch in the head of the pancreas (*bold arrow*).

b. MR virtual pancreatoscopy shows an internal view of the main pancreatic duct. The opening of the dilated side branch into the main pancreatic duct is visible (*arrow*).

c. MR virtual pancreatoscopy shows an internal view of the dilated side branch. The septal structure is depicted (*arrow*).

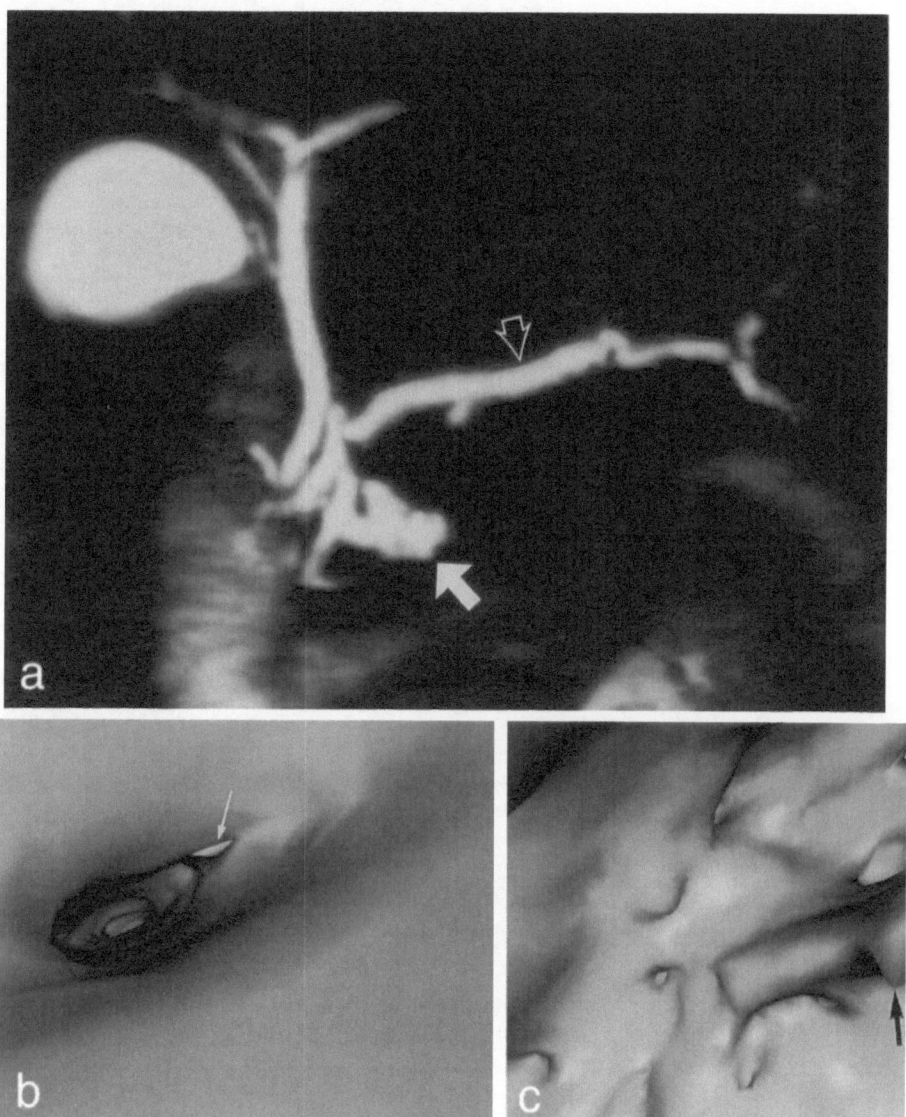

Fig. 6-8 Carcinoma of the papilla of Vater

a. Three-dimensional MRCP with MIP shows a filling defect at the ampulla (*arrow*).

b. MR virtual cholangioscopy shows a tumor at the ampulla (*bold arrow*). A PTCD tube (*open arrow*) is visible. An incision is being made from the bile duct toward the pancreatic duct (*fine arrow*).

c. MR virtual cholangioscopy shows a tumor at the ampulla (*bold arrow*) and the pancreatic duct (*fine arrow*).

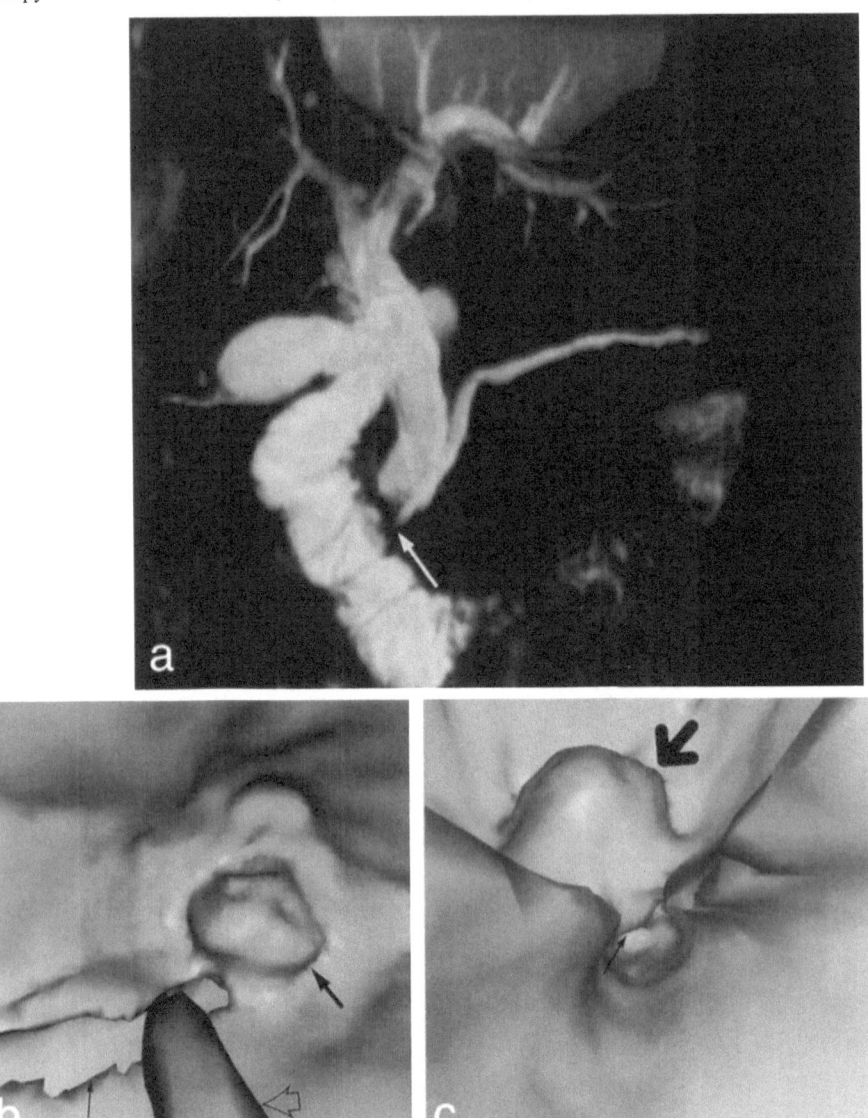

References

1. Takehara Y, Ichijo K, Tooyama N, et al (1995) Enhanced delineation of the pancreatic duct in MR cholangiopancreatography (MRCP) with combined use of secretin (in Japanese). Nippon Acta Radiologica 55:255-256

2. Takehara Y (1996) MR pancreatography: technique and applications. Top Magn Reson Imag 8:290-301

3. Matos C, Metens T, Deviere J, et al (1997) Pancreatic duct: morphologic and functional evaluation with dynamic MR pancreatography after secretin stimulation. Radiology 203:435-441

4. Sai J, Ariyama J, Suyama M, et al (1997) MRCP-secretin test (in Japanese). J Jpn Pancr Soc 12:193

5. Neri E, Boraschi P, Braccini G, et al (1999) MR virtual endoscopy of the pancreatobiliary tract. Magn Reson Imag 17:59-67

Index